ESSENTIAL ANIMAL BEHAVIOR

For Lisa, William, and Adam

Essential Animal Behavior

Graham Scott

Department of Biological Sciences, University of Hull, Hull, UK

Blackwell
Publishing

©2005 by Blackwell Science Ltd
a Blackwell Publishing company

BLACKWELL PUBLISHING
350 Main Street, Malden, MA 02148-5020, USA
9600 Garsington Road, Oxford OX4 2DQ, UK
550 Swanston Street, Carlton, Victoria 3053, Australia

The right of Graham Scott to be identified as the Author of this Work has been asserted in accordance with the UK Copyright, Designs, and Patents Act 1988.

First published 2005 by Blackwell Publishing Ltd

6   2009

Library of Congress Cataloging-in-Publication Data

Scott, Graham (Graham W.)
    Essential animal behavior / Graham Scott.
        p.   cm.
Includes bibliographical references (p.).
ISBN: 978-0-632-05799-3 (pbk. : alk. paper)
1. Animal behavior. I. Title.

QL751.S362004
591.5-dc22

                                    2003023583

A catalogue record for this title is available from the British Library.

Set in 10/13pt Stone Serif
by Graphicraft Ltd, Hongkong
Printed and bound in Singapore
by Ho Printing Singapore Pte Ltd

The publisher's policy is to use permanent paper from mills that operate a sustainable forestry policy, and which has been manufactured from pulp processed using acid-free and elementary chlorine-free practices. Furthermore, the publisher ensures that the text paper and cover board used have met acceptable environmental accreditation standards.

For further information on
Blackwell Publishing, visit our website:
www.blackwellpublishing.com

Contents

Boxes

Preface

My aim in writing this book has been to provide a concise but thorough introduction to the study of animal behavior. I want to convey the idea that animal behavior is a multidisciplinary field which draws into it many aspects of the broader field of biology.

In an introductory level textbook, it is impossible to cover all of the classic works of earlier times, but I have incorporated a sufficient number of them to provide the reader with some sense of the development of the field. I have also provided information on very recent and current work, and indicated that some questions remain unanswered.

My primary goals have been to produce a book that will be readable, be useful to both students and tutors, and will encourage readers to pursue their interest further.

The book puts the study of animal behavior in an applied context, emphasizing the implications for animal welfare and animal conservation. Social behavior is covered throughout and new, exciting examples from both the terrestrial and marine environments highlight current research alongside the classic examples.

Aimed at undergraduate students taking introductory and non-majors courses in animal behavior and related areas, the book is essential reading for degree-level students in biology, zoology, marine biology, and psychology departments.

Various pedagogical features have been incorporated into the book, and it has been carefully designed to meet the needs of students studying the subject for the first time. The features included are explained below:

- **"Focus on" boxes:** cover selected points, examples, and concepts in more depth.
- **"Concept" boxes:** highlight key terms and concepts, allowing students to see at a glance the important themes and ideas covered in each chapter.
- "Application" boxes: describe how theory can be applied to real-world examples, bringing the subject to life.
- **"Case study" boxes:** take examples a step further, providing extra information and allowing more in-depth discussion.

Other features include:
- **Chapter summaries:** aid understanding, provide a quick reference to each chapter, and can help to guide revision.
- **"Questions for discussion" boxes:** encourage students to think in more depth about key topics and provide discussion points for tutorials.
- **"Key reference" boxes and Further reading:** allow the student to take each topic further, highlighting key papers and good sources of information, and helping to guide revision.
- **"Link" boxes:** color-coded links to other chapters in the book provide cross-referencing between related areas. These help students navigate around the book and also serve to demonstrate the interrelationships between the topics.

Additional resources for lecturers are available either as a CD ROM or for download from www.blackwellpublishing.com/scott. These include all the figures and artwork in Powerpoint and in JPEG format.

Acknowledgments

Although the book has a single author, it is not the result of the efforts of a single person. Without the dedication of those persons actively engaged in animal behavior research I would have nothing to write about! They further our understanding of the world and deserve our thanks for it. I would like to thank my family for putting up with me during the writing of the book, and especially Lisa for all the help that she gave me. The editorial and publishing team that have produced the book have given me gratefully received support. Ian Sherman, Sue Hull, Michelle Tobin, Chris Saunders, Magnus Johnson, and Ian McFarlane all patiently read and commented on various bits of the text, and I am indebted to them all.

The author and publisher gratefully acknowledge the permission granted to reproduce the copyright material in this book.

Fig. 1.2: Tinbergen, N. (1963) The shell menace. *Natural History*, **72**, 28–35.

Fig. 2.3: Carew, T.J. (2000) *Behavioral Neurobiology*. Sinauer Associates, Sunderland, MA.

Fig. 2.4: Carew, T.J. (2000) *Behavioral Neurobiology*. Sinauer Associates, Sunderland, MA.

Fig. 2.5: Barth, F.G. & Höller, A. (1999) Dynamics of arthropod filiform hairs. V. The response of spider trichobothria to natural stimuli. *Philosophical Transactions of the Royal Society of London, Series B*, **354**, 183–92.

Fig. 2.6: Ewart, J-P. (1985) Concepts in vertebrate neuroethology. *Animal Behaviour*, **33**, 1–29.

Fig. 2.7(a): Camhi, J.M. (1984) *Neuroethology: Nerve Cells and the Natural Behaviour of Animals*, p. 231, Sinauer Associates, Sunderland, MA; from Muntz, W.R.A. (1964) Vision in frogs. *Scientific American* **210**(3), 110–19.

Fig. 2.7(b): Camhi, J.M. (1984) *Neuroethology: Nerve Cells and the Natural Behaviour of Animals*, p. 231, Sinauer Associates, Sunderland, MA; from Ewart, J.P. (1980) Concepts in vertebrate neuroethology. *Animal Behaviour*, **33**, 1–29.

Fig. 2.9: Ewart, J-P. & von Wietershein, A. (1974) Pattern analysis by tectal and thalamus/pretectal nerve nets in the visual system of the toad *Bufo bufo* (L.). *Journal of Comparative Physiology*, **92**, 131–48.

Fig. 2.10: Wine, J. & Kranse, F.B. (1972) The organisation of escape behaviour in the crayfish. *Journal of Experimental Behaviour*, **56**, 1–18.

Fig. 2.11: Wine, J. & Kranse, F.B. (1982) The cellular organisation of crayfish escape behaviour. In *The Biology of Crustacea*, ed. by Bliss, D.E. *et al.*, pp. 241–92, Academic Press, London.

Fig. 2.12: Carew, T.J. (2000) *Behavioral Neurobiology*. Sinauer Associates, Sunderland, MA.

Fig. 3.1: Bull, C.D. & Metcalfe, N.B. (1997) Regulation of hyperphagia in response to varying energy deficits in overwintering juvenile Atlantic salmon. *Journal of Fish Biology*, **50**, 498–510.

Fig. 3.2: Sánchez-Vázquez, F.J. & Tabata, M. (1998) Circadian rhythms of demand-feeding and locomotor activity in rainbow trout. *Journal of Fish Biology*, **52**, 255–67.

Fig. 3.3: Morgan, E. & Cordiner, S. (1994) Entrainment of circa-tidal rhythm in the rock-pool blenny *Lipophrys pholis* by simulated wave action. *Animal Behaviour*, **47**, 663–9.

Fig. 3.4: Morgan, E. & Cordiner, S. (1994) Entrainment of circa-tidal rhythm in the rock-pool blenny *Lipophrys pholis* by simulated wave action. *Animal Behaviour*, **47**, 663–9.

Fig. 3.5: Toates, F., ed. (1998) *Control of Behaviour*. Springer-Verlag, Berlin.

Fig. 4.1: DeBelle, J.S. & Sokolowski, M.B. (1987) Heredity of rover/sitter: alternative foraging strategies of *Drosophila melanogaster* larvae. *Heredity* **59**, 73–83.

Fig. 4.2: Hall, M. & Halliday, T., eds (1998) *Behaviour and Evolution*. The Open University/ Springer Verlag, Berlin.

Fig. 4.3: Hall, M. & Halliday, T., eds (1998) *Behaviour and Evolution*. The Open University/ Springer Verlag, Berlin; from Bentley, D. & Hoy, R.R. (1972) Genetic control of the neuronal networks generating cricket song patterns. *Animal Behaviour*, **20**, 478–92.

Fig. 4.4: Hailman, J.P. (1969) How an instinct is learned. *Scientific American*, **221**(6), 106.

Fig. 4.5: Carew, T.J. (2000) *Behavioral Neurobiology*. Sinauer Associates, Sunderland, MA; from Wine, J. (1975) Habituation and inhibition of the crayfish lateral giant fibre escape response. *Journal of Experimental Biology*, **62**, 771–82.

Fig. 4.6: Kandel, E.R. (1979) Small systems of neurons. *Scientific American*, **241**(3), 67–76.

Fig. 4.7: Kandel, E.R. & Schwartz, J.H. (1982) Molecular biology of learning: modulation of transmitter release. *Science*, **218**, 433–43.

Fig. 4.8: Kandel, E.R. (1984) Steps towards a molecular grammar for learning: explorations into the nature of memory. In *Medicine, Science and Society*, ed. by Isselbacher, K.J., pp. 555–604, Wiley, New York; from Carew, T.J., Walters, E.T. & Kandel, E.R. (1981) Classical conditioning in a simple withdrawal reflex in *Aplysia californica*. *Journal of Neuroscience*, **1**, 1426–37.

Fig. 4.9: Pearce, J.M. (1997) *Animal Learning and Cognition: An Introduction*. Psychology Press, Hove.

Fig. 4.10: Pearce, J.M. (1997) *Animal Learning and Cognition: An Introduction*. Psychology Press, Hove; from Macphail, E.M. (1993) *The Neuroscience of Animal Intelligence*. Colombia University Press, New York.

Fig. 4.11: Griffin, A.S. *et al.* (2001) Learning specificity in aquired predator recognition. *Animal Behaviour*, **62**, 577–89.

Fig. 4.12: Witherington, B. (1997) The problem of photopollution for sea turtles and other nocturnal animals. In *Behavioural Approaches to Conservation in the Wild*, ed. by Clemmons, J.R. & Buchholz, R., pp. 303–28. Cambridge University Press, Cambridge.

Fig. 4.13: Shen, J.X. *et al.* (1998) Direct homing behaviour in the ant *Tetramorium caespitum* (Formicidae, Myricinae). *Animal Behaviour*, **55**, 1443–50.

Fig. 4.14: Fukushi, T. (2001) Homing in wood ants, *Fornica japonica*, use of a skyline panorama. *Journal of Experimental Biology*, **204**, 2063–72.

Fig. 4.15: Cheng, K. & Spetch, M.L. (1998) Mechanisms of landmark use in mammals and birds. In *Spatial Representation in Animals*, ed.

by Healy, S., pp. 1–17. Oxford University Press, Oxford.

Fig. 4.16: Morris, R.G.M. (1981) Spatial localization docs not require the presence of local cue. *Learning and Motivation*, **12**, 239–60.

Fig. 4.17: Healey, S.D., Clayton, N.S. & Krebs, J.R. (1994) Development of hippocampal specialisation in two species of tit (*Parus* spp.). *Behavioural Brain Research*, **81**, 61–8.

Fig. 4.19: Alcock, J. (2001) *Animal Behaviour, An Evolutionary Approach*. Sinauer Associates, Sunderland, MA; from Helbig, A.J. (1991) Inheritance of migratory direction in a bird species: a cross-breeding experiment with SE- and SW-migrating black caps (*Sylvia atricapilla*). *Behavioural Ecology and Sociobiology*, **42**, 9–12.

Fig. 5.2: Rosenthal, G.G. & Evans, C.S. (1998) Female preference for swords in *Xiphophorus helleri* reflects a bias for large apparent size. *Proceedings of the National Acdaemy of Science, USA*, **95**, 4431–6.

Fig. 5.3: Rosenthal, G.G. & Evans, C.S. (1998) Female preference for swords in *Xiphophorus helleri* reflects a bias for large apparent size. *Proceedings of the National Acdaemy of Science, USA*, **95**, 4431–6.

Fig. 5.5: de la Torre, S. & Snowdon, C.T. (2000) Environmental correlates of vocal communication of wild pygmy marmosets, *Cebulla pygmaea*. *Animal Behaviour*, **63**, 847–56.

Fig. 5.6: de la Torre, S. & Snowdon, C.T. (2000) Environmental correlates of vocal communication of wild pygmy marmosets, *Cebulla pygmaea*. *Animal Behaviour*, **63**, 847–56.

Fig. 5.7: Marler, P. (1959) Developments in the study of animal communication. In *Darwin's Biological Work*, ed. by Bell, P.R., pp. 150–202. Cambridge University Press, Cambridge.

Fig. 5.8: Zuberbühler, K. (2002) A syntactic rule in forest monkey communication. *Animal Behaviour*, **63**, 293–9.

Fig. 6.1: Cowan, D.P. *et al.* (2000) Reducing predation through conditioned taste aversion. In *Behaviour and Conservation*, ed. by Gosling, L.M. & Sutherland, W.J., pp. 281–99. Cambridge University Press, Cambridge.

Fig. 6.2: Vissalberghi, E. & Addessi, E. (2000) Seeing group members eating a familiar food enhances the acceptance of novel foods in Capuchin monkeys. *Animal Behaviour*, **60**, 69–76.

Fig. 6.3: Green, E. (1987) Individuals in an osprey colony discriminate between high and low quality information. *Nature*, **329**, 239–41.

Fig. 6.4: Major, P.F. (1978) Predator–prey interactions in two schooling fishes, *Carex ignoblis* and *Stolephorus purpureus*. *Animal Behaviour*, **26**, 760–77.

Fig. 6.5: Göttmark, F., Winkler, D., Andersson, M. (1986) Flock-feeding on fish schools increases success in gulls. *Nature*, **319**, 589–91.

Fig. 6.6: Elner, R.W. & Hughes, R.N. (1978) Energy maximisation in the diet of the shore crab *Carcinus maenas*. *Journal of Animal Ecology*, **47**, 103–6.

Fig. 6.7: Kacelnik, A. (1984) Central place foraging in starlings (*Sturnus vulgaris*). I. Patch residence time. *Journal of Animal Ecology*, **53**, 283–99.

Fig. 6.8: Walton, P., Ruxton, G.D. & Pitelka, F.A. (1998) Avian diving, respiratory physiology and the marginal value theorem. *Animal Behaviour*, **56**, 165–74.

Fig. 6.10: Magnhagen, C. (1990) Conflicting demands in gobies: when to eat, reproduce, and avoid predators. In *Behavioural Ecology of Fishes*, ed. by Huntingford, F.A. & Toricelli, P., pp. 79–90. Harwood Academic Publishers, Oxford.

Fig. 7.1: Götmark, F. & Olsson, J. (1997) Artificial colour mutation: do red-painted great tits experience increased or decreased predation? *Animal Behaviour*, **53**, 83–91.

Fig. 7.2: Cresswell, W. (1994) Flocking is an effective anti-predator strategy in redshanks, *Tringia totanus. Animal Behaviour*, **47**, 433–42.

Fig. 7.3: Myers, J.P, Connors, P.G. & Pitelka, F.A. (1979) Territory size in wintering sanderlings: the effects of prey abundance and intruder density. *Auk*, **96**, 551–61.

Fig. 7.4: Krebs, J.R. & Davies, N.B. (1993) *An Introduction to Behavioral Ecology*. Blackwell Science, Oxford.

Fig. 7.5: Kruuk, H. (1964) Predators and anti-predator behaviour of the black-headed gull *Larus ridibundus. Behaviour Supplements*, **11**, 1–129.

Fig. 7.6: Brown, C. & Hoogland, J.L. (1986) Risk in mobbing for solitary and colonial swallows. *Animal Behaviour*, **34**, 1319–23.

Fig. 7.7: Hanlon, R.T. & Messenger, J.B. (1996) *Cephalopod Behaviour*. Cambridge University Press, Cambridge; based on Hanlon, R.T. & Messenger, J.B. (1988) Adaptive colouration in young cuttlefish (*Sepia officinalis* L.): the morphology and development of body patterns and their relation to behaviour. *Philosophical Transactions of the Royal Society, Series B*, **320**, 437–87.

Fig. 7.8: Caro, T.M. (1986) The functions of stotting in Thompson's gazelles: some tests of predictions. *Animal Behaviour*, **34**, 663–84.

Fig. 7.9: Caro, T.M. (1986) The functions of stotting in Thompson's gazelles: some tests of predictions. *Animal Behaviour*, **34**, 663–84.

Fig. 8.1: Amundsen, T. & Forsgren, E. (2001) Male mate choice selects for female colouration in a fish. *Proceedings of the National Academy of Sciences of the USA*, **98**, 13155–60.

Fig. 8.2: Moczek, A.P. & Emlen, D.J. (2000) Male horn dimorphism in the scarab beetle, *Onthophagus tarsus*: do alternative reproductive tactics favour alternative phenotypes? *Animal Behaviour*, **59**, 459–66.

Fig. 8.3: Davies, N.B. (1992) *Dunnock Behaviour and Social Evolution*. Oxford University Press, Oxford.

Fig. 8.4: Jones, J.S. & Wynne-Edwards, K.E. (2001) Paternal behaviour in biparietal hamsters, *Phodopus campbelli*, does not require contact with the pregnant female. *Animal Behaviour*, **62**, 453–64.

Fig. 8.5: Davies, N.B. (1992) *Dunnock Behaviour and Social Evolution*. Oxford University Press, Oxford.

Fig. 8.6: Jenkins, E. *et al.* (2000) Delayed benefits of paternal care in the burying beetle *Nicroplorous vespilloides. Animal Behaviour*, **60**, 443–51.

Fig. 8.8: Alatalo, R.V. & Lundberg, A. (1984) Polyterritorial polygyny in the pied fly-catcher *Ficedula hypoleuca. Annales Zoologici Fennici*, **21**, 217–28.

Fig. 8.9: Krebs, J.R. & Davies, N.B. (1993) *An Introduction to Behavioural Ecology*. Blackwell Science, Oxford; from Orions, G.H. (1969) On the evolution of mating systems in birds and mammals. *American Naturalist*, **104**, 589–603.

Fig. 8.10: Shelley, T.E. (2001) Lek size and female visitation in two species of tephritid fruit flies. *Animal Behaviour*, **62**, 33–40.

Table 4.1: Arathi, H.S. & Spivak, M. (2001) Influence of colony genotypic composition on the performance of hygienic behaviour in the honey bee (*Apis mellifera* L.). *Animal Behaviour*, **62**, 57–66.

Every effort has been made to trace copyright holders and to obtain their permission for the use of copyright material. The publisher apologizes for any errors or omissions, and would be grateful to be notified of any corrections that should be incorporated in future reprints or editions of this book.

1 Essential Animal Behavior: An Introduction

Frequently consider the connection of all things in the universe and their relation to one another.

Marcus Aurelius AD 121–80

People have probably always been fascinated by the behavior of animals. Indeed an understanding of the behavior of prey animals must have been essential to our early ancestors; their paintings on the walls of caves suggest that they could have been fairly familiar with behavioral concepts such as herd size and migration. The earliest stock-farmers would have needed to understand the behavior of the charges in their care just as their modern counterparts do today.

Some members of society (and even some biology students) may wrongly think of the study of animal behavior in an academic context as being a soft science or even an easy option.

Contents

Behavior
Anthropomorphism
Questions about
Causation
Evolution
Function
Ontogeny
Adaptation
Applications
Animal welfare
Conservation
Summary
Questions for discussion

Key points

◆ The field of animal behavior is diverse and may be studied from a variety of perspectives.

◆ It is useful to consider behaviors as adaptations.

◆ A single behavior will not serve, or serve the same purpose in all situations, and behaviors are adapted to be effective in the environment of the animal performing them.

◆ It is wrong to think of animal behavior as a general interest or a purely academic subject. The study of animal behavior is an important science which has a clear applied context.

However, I hope to show you in this introduction to the subject that it is an important and rigorous science and that it has a clear application to some of the problems that we face in the modern world.

Cephalopod inking behavior

Many species of octopus and squid are known to exhibit a particularly effective behavior that enables them to escape from predators. In the region of their intestines the animals have a special sac-like organ. In the wall of this sac there is a gland which secretes a brown or black liquid rich in the pigment melanin, this is ink. When threatened the animal has the ability to compress the ink sac and squirt a jet of the liquid from its anus. It is thought that the cloud of ink hanging in the water forms a dummy squid termed a pseudomorph, which attracts and holds the attention of the predator allowing the animal to dart away to safety. The deception is made all the more effective because long thin species produce long thin pseudomorphs and more round species produce rounder clouds of ink (Plate 1.1).

Squid and octopus are molluscs, taxonomic relatives of the garden slug and snail. Can you imagine a slug squirting out ink to leave a pseudomorph hanging in the air to decoy a bird predator while the slug made its escape? Of course you can't, for the simple reason that this behavioral strategy can only work when the animal is surrounded by a medium that will support the ink cloud for a sufficient period to allow the escape. In water this works, but in the less dense medium of air it would not.

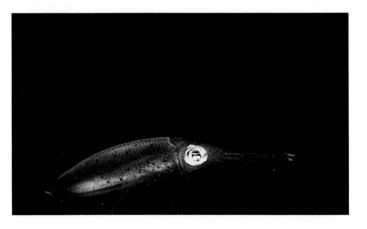

Plate 1.1 An animal this shape should produce a long, thin pseudomorph. © C. Waller.

Some species of octopus and squid are inhabitants of the ocean depths. Here light penetration from the surface is minimal or zero and the seawater is a constant inky black. Obviously the ink-dummy strategy would be no more effective here than it would be in air. The pseudomorph would hang in the water column, but it is unlikely that an ink-black shape would be seen against the inky-black backdrop. In this situation species such as the deep-water squid *Heteroteuthis* secrete a luminescent ink, creating a brief flash of light which is thought to confuse a potential predator just long enough for an escape to be affected.

From this example I hope that I have made a few key points about behavior. Firstly, that behaviors are adaptations which serve specific functions, and we will consider this point further later in this chapter. Secondly, that a single behavior may not serve, or serve the same function, in all situations (a point to be borne in mind throughout this book). Finally, behaviors are adapted to be effective in the environment of the animal performing them.

What is behavior?

Before investigating the amazing diversity of behaviors that animals exhibit, it is necessary for us to gain some insight into the concept of behavior itself. We need to decide what the word **behavior** means to us in the current context and to examine the various avenues open to us for the study of animal behavior.

So what is behavior? Dictionary definitions of the word typically include phrases such as "acting or functioning in a specified or usual way." This suggests to us that behavior is a predictable thing. Another common phrase is "the response of an organism to a stimulus." This suggests that behaviors are **made** to happen by something. In the case of this definition the "something" concerned is not specified, and may be internal or external to the animal involved. Each of these ideas is in its own way an adequate response to the question. Behaviors are in many cases predictable given sufficient information concerning their context (although many appear initially to be highly unpredictable). Similarly behaviors are often linked to a stimulus in an immediate sense at some level. The shortcoming of such definitions, however, is that they attempt to narrowly confine behavior in an easily described and highly specific way. Given the diversity of behavior such an approach may not be appropriate, because as humans we often think of different behaviors in very different terms.

Take for example breathing, swimming, and learning. Each of these words describes a behavior for which the definitions presented above would be sufficient. However as humans we would probably not think of them as being equivalent conceptually. We would consider breathing to be an involuntary process and may not even consider it to be a behavior at all because of that. Swimming, on the other hand, is clearly an active process, we tend to think of it as having a motivation or goal. Learning we think of in different terms again. We have a tendency to place it into a higher class of processes, which require a higher level of mental ability (though we shall see in Chapter 4 that this need not be the case). So as a result of our own preconceptions about the words we use to label behaviors, and their obvious diversity, it will be much more useful for us to adopt a very broad definition of behavior in this book.

Put more simply then behavior is a property of all living things and whenever we observe an animal to be engaged in any activity (voluntary or involuntary) we are witnessing behavior. Indeed it could be said that the only animal not behaving is a dead animal!

This is an important point to remember. Although we may feel that a sleeping seal or a motionless sea-snake are doing nothing, they are in fact behaving. The act of sleep is a behavior in its own right, and the snake is quite probably poised to strike at passing prey, tensing a host of muscles in readiness and taking in and processing a wealth of information about its environment. There can be no doubt that it is behaving.

Focus on anthropomorphism

Anthropomorphism is the attribution of human feelings and emotional states to animals. As humans we are aware of three mental experiences: feelings (pain, pleasure, etc.), motivations (the purposes of our actions), and thought. Our current understanding of nonhuman animal species does not allow us to say that they experience the same (of course nor does it preclude common mental experiences). Throughout the development of the field of animal behavior anthropomorphism and the use of anthropomorphic language has been frowned upon by some as a bar to clarity of expression, and in the extreme as a bar to the progression of science. However, given the "human baggage" associated with many of the words we use in describing behavior (e.g. aggression, hierarchy, motivation), it is inevitable that a degree of anthropomorphism will occur.

Throughout this text I have endeavored to avoid unwarranted anthropomorphism; in reading it I would ask that you do the same.

Approaches to the study of animal behavior

The effective study of animal behavior requires observation and experimentation. Before we have any chance of understanding a behavior fully we must observe that behavior, in its natural context, and in its entirety. The careful description of a behavior pattern or a sequence of behaviors allows us to identify all of the relevant components and to link their performance to the wider context of the physical and biological environment of the animal. From such a knowledge base we are able to develop our own ideas about that behavior, to speculate upon its function, and upon the factors which control it. In the language of science we are able to generate testable hypotheses. These are carefully worded questions that we hope to answer via carefully designed experiments where specific factors surrounding behavioral performances are monitored and manipulated. Throughout this book we will consider the results of such observations and experiments.

Animal behavior can be studied at two key levels. At the physiological level we might be interested in the mechanism by which a behavior actually occurs. By this I mean in what way do the biochemistry, nerves, muscles, and senses of an animal interact to result in a particular behavior? Equally interesting, however, are questions related to the whole animal and the world external to it. At this level we might consider the performance of a behavior in relation to the environment in which it is performed, to the wider ecology of the animal, or to its social experiences. This kind of whole animal observation could be carried out in the field – in the natural environment of the animal – or it could be carried out in the laboratory where a controlled environment more readily permits experimentation. In recent times, however, these distinctions have blurred and it is now commonplace to have "field" simulations in the laboratory and experimental manipulations in the field. The most significant recent development in the process of the study of animal behavior must however be the advent of powerful user-friendly computers. Via computer models behavior can be simulated, and "experiments" carried out and evaluated without the involvement of any animal (other than the human operator) and often at a fraction of the cost of a more traditional line of investigation. Such an approach undoubtedly has value, but a model is only as good as the data provided to it, and it will never replace the study of living animals.

Asking questions in the study of behavior

As has been previously mentioned, we further the study of animal behavior by posing (and hopefully answering) questions,

> **Concept**
> **The scientific method**
>
> The scientific method is a four-stage approach that we can use to explore animal behavior.
>
> First we make observations of a behavioral phemonenon. Then we use these observations to formulate an hypothesis to explain the behavior. This may lead us to make further predictions concerning the behavior. Finally we design, carry out, and evaluate experiments to test our hypothesis and predictions.

Link
Models enable us to fully explore behavior.
Chapter 6

which are carefully constructed to take into account previous observations (and the answers to previous questions). Although the actual number of specific questions that we can ask is huge, there are a relatively small number of types of question that need concern us:

- **What is behavior X?** The simplest way to answer this question would be to provide a description of the behavior. A more sophisticated answer might consider the raison d'être of the behavior, and in doing so the question would overlap with those below.
- **When does behavior X occur?** We might address this question in a number of ways. Perhaps we are asking at what time of year/day does it occur? Alternatively the question could relate to a life stage (is the behavior performed by mature animals only?). We could also be interested in the position of this behavior within a behavioral sequence (does the animal always do X after Y? or does one animal always do X if another animal has just done Z?).
- **Why does behavior X occur?** This is perhaps the most often asked question during studies of animal behavior and it is certainly the question which is given the most weight.

Tinbergen's four questions

In 1963 Niko Tinbergen, a recognized pioneer of the study of animal behavior, suggested that there are four ways in which we can ask the question "why?" Such is the importance afforded to Tinbergen's four questions (as they are usually referred to) that it is worth spending some time on them so that we can be sure that we understand the subtleties of the distinctions between these four ways of asking "why?" It is also important to remember that no one of the possible answers to the question "why?" that we might discuss is more important than any other. They complement one another and together help us to appreciate the wider picture.

Why do cephalopods ink?

Key reference
Tinbergen, N. (1963) On aims and methods of ethology. *Zeitschrift für Tierpsychologie*, **20**, 410–33.

We started this chapter with a description of the way in which some species of cephalopod mollusc eject a cloud of ink as part of their antipredator behavior. As enquiring animal behaviorists we should not be satisfied with a description of the behavior, we should ask the question why does it happen? Tinbergen suggested that to fully explore the behavior we should ask **why**? in terms of the **causation** of the behavior, in terms of its **evolution**, of its **function**, and of its **ontogeny**.

Causation

What causes a cephalopod to ink? The highly developed sensory organs of the cephalopods allow an animal to continually receive a wealth of information regarding its environment. The animal processes this information and specific environmental stimuli elicit specific responses. If the animal is threatened with immediate danger a sequence of nerve impulses trigger the activity of the ink sac, an effector organ (i.e. an organ which carries out the response to a stimulus). Upon stimulation the ink sac compresses and the duct sealing it from the rectum opens. A jet of ink is expelled through the rectum and out of the anus. In the environment the ink forms either a pseudomorph or a diffuse cloud, depending upon the species involved.

In thinking about the roles of sense organs, nerves, and muscles we have already considered the mechanics of inking behavior, and so we have begun to examine the causal mechanisms of this behavior in an immediate sense. We might go further, however, to better understand the role of the recent experiences of the animal in the process. Inking is relatively costly because it involves the production of an energetically expensive substance. For this reason it is usually a behavior of last resort. Typically an individual may have attempted to remain undetected through camouflage (crypsis behavior) or to frighten or confuse its attacker by performing one of a range of diematic behaviors. Examples of diematic behavior include changing color rapidly to startle a predator or adopting a threatening body posture. We should therefore consider failure of crypsis and diematic behavior as being causal factors of inking.

Evolution

Behaviors do not fossilize in the same way that body parts do. For this reason we can say little about the very recent evolution of inking behavior, or about its continuing development in the face of improvements in the ability of predators to detect the bluff, as must surely happen. However, by looking at the family tree of the molluscs, and at that of the cephalopods particularly, we can say something about the longer-term evolution of the behavior. Inking is not a general property of the molluscs, it is restricted to

Link
Behaviors are coordinated by the actions of nerves and muscles.
Chapter 2

Link
An animal may employ a range of different antipredator behaviors.
Chapter 7

the class Cephalopoda. This tells us that the behavior has evolved more recently than at the point at which the cephalopods diverged from the other molluscs. Further investigation reveals that not all cephalopods have an ink sac. The order is subdivided into two subclasses, the Nautilioidea (none of the species of nautilus has an ink sac) and the Coleoidea (most of which have an ink sac). Inking must thus be a relatively recent evolution within the subclass Coleoidea, the squids and octopuses. Further examinations of differences and similarities in inking behavior between the various octopus and squid families would allow us to gain yet more insights into the evolution of the behavior, and may help us to better understand why it has evolved.

Function

The function of inking behavior is very easy to understand. The behavior allows an animal to escape, thereby ensuring its survival and contributing to its fitness. (Fitness in this context equates to reproductive fitness and is a measure of the organism's success in passing on its genes – after all a dead animal cannot pass on its genes.) If inking is so effective why do some octopus and squid use it only as a last resort? As we have already mentioned inking probably has a high cost associated with it, whereas the behaviors involved in crypsis are probably far less costly. It would be better for an individual to rely on less costly behaviors where possible, because by doing so it may save energy with which its fitness might be enhanced. For example, we could speculate that by not using and replacing ink an animal may be able to spend more energy producing a greater number of eggs or sperm.

Ontogeny

We might expect such a useful behavior as inking to be performed by all animals throughout their lives. After all the cephalopods have many predators. However, this is not always the case. The young of the blue-ringed octopus (*Hapalochlaena lunulata*) eject ink during the first few weeks of life, but after that they never do. *H. lunulata* is highly venomous – perhaps this proves a sufficient deterrent to predators of adults, but not to those preying upon the young animals. Antipredator behavior must therefore continue to develop and be refined throughout the lifetime of the animal in at least some species. The cover of this book shows just how striking the blue-ringed octopus is.

Link
The modification of behavior may continue throughout the life of an individual.
Chapter 3

Behaviors as adaptations

As biologists we are used to talking about adaptations such as the development of antibiotic resistance in bacteria, or about organs like feathers and wings as adaptations for flying, but we perhaps do not think about behaviors as adaptations as often as we should.

Imagine for example a population of squid at a time early in the evolution of inking behavior. Faced with predators our hypothetical squid might have been able to do no more than squirt a diffuse cloud of ink, which dissipated in the water relatively quickly and had limited success (but some success) as an antipredator behavior. Now imagine that a genetic mutation occurred in a single animal, which had the effect that a chemical in the ink was altered slightly in its composition so that it became gooey on contact with water. When this animal squirted, the ink formed a rudimentary pseudomorph rather than a cloud. What would have been the effect of this alteration in antipredator behavior? Well perhaps it increased the likelihood of survival for this one animal.

If, however, the mutation was a heritable one that would be passed on to all of the young of that animal then something very important indeed might occur. As a result of its increased survival probability this animal might enjoy better reproductive success than the average squid, and because they inherited its gooey ink gene its offspring might also

> ### Focus on **heredity and natural selection**
>
> Any student of biology who takes the time to think about the characteristics of populations of individuals should note the following:
> - That there is considerable variation among the individuals within a species.
> - That this variation is passed from parents to their offspring through their genes.
> - That many more individuals are produced than can ever survive to mature and reproduce.
>
> From these basic building blocks, together with a prodigious amount of patient work and probably a touch of genius, Charles Darwin was able to construct his theory of evolution by natural selection. Long before the importance of genes was appreciated he realized that some heritable property of certain individuals placed them at an advantage relative to others of their kind. Through time the differential survival of these individuals could result in a shift in the make-up of the population such that animals without the advantage became increasingly rare (or perhaps disappeared) and those possessing it became more common. He proposed a scenario under which beneficial traits were selected for and detrimental traits were selected against (Fig. 1.1).

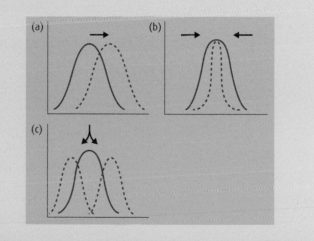

Fig. 1.1 An illustration of the effect of directional (a), stabilizing (b), and disruptive (c) selection upon a hypothetical population. In each case the solid line describes the pre-selection population and the dashed line the post-selection population.

We now understand that selection can be strongly directional in the way that I have just described, leading to the increasing dominance of a trait or to its eradication, or it can act in a disruptive way resulting in the evolution of a population that has two very different but successful phenotypes. Selection can also act in a stabilizing way to maintain the status quo.

Link
There is a genetic component to the performance of behavior.
Chapter 4

enjoy higher reproductive success than average. Through the course of a number of generations animals without the gene would become steadily less common whilst those with the gene would become more common, i.e. natural selection would be taking place. The animals better able to pass on their genes are being selected for, and those that suffer lower reproductive success are being selected against. The end result is the development of pseudomorph production as an adaptation to avoid predation.

Of course this is a purely hypothetical example, and following the course of development of such an adaptation through evolutionary time would hardly be possible. However there have been a number of studies that attempt to use real animals in their natural environments to better understand behavioral adaptation.

Perhaps the most widely quoted of these involves the work of Niko Tinbergen as part of his research into the behavior of various species of seabird. In the course of observations of a breeding colony of black-headed gulls (*Larus ridibundus*) he noticed two things. Firstly that the eggs in a nest do not all hatch on the same day, and secondly that parent gulls habitually remove the broken shells of their recently hatched eggs and deposit them some distance from their breeding area. In performing this behavior, however, it seemed to Tinbergen that the parent bird must place its newly hatched chick at risk because unattended chicks can and do fall easy prey to predatory birds including neighbouring gulls. So in 1967 Tinbergen asked the apparently simple question "why do black-headed gulls remove empty eggshells from their nests?" (Plate 1.2).

Tinbergen assumed (as I am sure would you) that the birds must remove the eggs because it was more advantageous to do so than to leave them in the nest. However, because all black-headed gulls perform this behavior it was not possible for him to simply compare the relative fitness of birds which did and did not remove eggshells. Instead he constructed an artificial colony of nests that he was able to manipulate. Some of the nests contained whole eggs only with some eggshells placed a fixed distance from the nest (to simulate nests from which the shell of the first hatched eggs had been removed). Others nests contained whole eggs and a broken eggshell (i.e. nests from which removal had

Plate 1.2 Black-headed gulls build their nests in close proximity to one another. © P. Dunn.

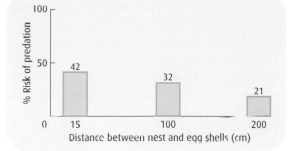

Fig. 1.2 The further away from the nest Tinbergen placed the eggshells, the lower the risk that the nest would be discovered by crows and the eggs eaten. (Data from Tinbergen, N. (1963) The shell menace. *Natural History*, 72, 28–35.)

not taken place). After a period of observation he found that the nests from which the eggshells had been removed were the most successful (and that the further the eggshells were taken from the nest the more successful was the strategy, see Fig. 1.2). Predators such as crows and other species of gull were far more likely to eat the eggs from the none-removal nests than from the removal nests. From this we could deduce that if a "mutation" in eggshell removal behavior were to occur in which parent birds did not remove, then selection (in the form of increased predator risk) would act against it. This is an excellent demonstration of the usefulness of the **experimental method** as a tool for the study of behavior.

If removal does confer a selective advantage then surely parent birds should remove the eggshells as soon as the chicks hatch rather than waiting for a short period of time prior to removal (which they are observed to do). Further experiments revealed the reason for this delay. Newly hatched chicks are wet and take a short time to dry out. During this period neighbouring black-headed gulls can easily swallow them. If the parent were to leave the nest to remove the eggshells during this crucial period, it is likely that the advantage of removal (the benefit) would be out-weighed by the disadvantage of cannibalism (the cost). This kind of reasoning is a vital tool for students of behavior and is termed a **cost–benefit analysis**.

Further confirmation of the role of predators as a selective agent in the shaping of eggshell removal behavior comes from the use of another key study method available to students of behavior. In 1957 Esther Cullen compared the behavior of the kittiwake (*Rissa tridactyla*) with that of a range of closely related gull species. She found that eggshell removal was a behavior common to the majority of gull species, but that it was never performed by the kittiwake. She suggested that the reason for this was that whereas all of the other gulls were ground-nesting birds who would be vulnerable to aerial predators such as crows, the nests of kittiwakes are constructed on the sides of high and sheer cliff faces and are not at risk. Thus this methodology, termed the **comparative approach** (because it involves comparing one species or group of species with another), also points to predation as a selective force in the adaptation of eggshell removal.

Throughout this book we will consider the design and the results of a number of studies similar to those I have described above. We will investigate the way in which the environment has played a part in the shaping of behavior and we will consider behaviors in an adaptive sense. By doing so we will gain new insights into the behavior of animals and into the process of behaving itself.

Why study behavior?

I opened this chapter with the observation that as a species we have a fascination with the behavior of animals (including our-selves). Personally I think that the "it's fascinating argument" is all the justification I need when friends and colleagues ask me why I study animal behavior, but I do regularly find myself having to provide other reasons when pressed. In fact the study of animal behavior has an application to a number of areas of modern life.

As a very obvious example, there is no doubt that an increased understanding of behavior does contribute to an increased understanding of ourselves as an "animal." Comprehension of the physical basis of human memory and brain function has certainly been advanced as a result of breakthroughs made using nonhuman animal models. The other main applications of behavior are however a little less obvious, and it is these areas that I want to draw to your attention in the remainder of this chapter and in the form of specific case studies throughout the remainder of this book.

Behavior and animal welfare

During the closing decades of the twentieth century large numbers of people became vociferous advocates for the welfare of animals. Initially the movement concerned itself with the welfare of captive animals; those species reared on farms, exhibited in zoos, and kept as companions by humans. More recently there has been a shift towards concern for the welfare of wild animals: those that are hunted or trapped, and those that are managed for conservation. In the case of the latter group a dilemma has presented itself as to when we should intervene and when we should "let nature take its course." Essentially the role of animal behavior in this area has been to play a part in the identification of examples of poor welfare, to explore and explain behavioral manifestations of poor welfare, and to provide potential solutions.

Key reference
Spedding, C. (2000) *Animal Welfare*. Earthscan Publications, London.

At a simple level we can think of welfare as equating to a state of well being or some form of contentment, and lack of obvious suffering as a result of cruelty. But welfare is more than that. An animal that enjoys good welfare will not want for food, water, or shelter. It will have space to move and opportunities for socialization appropriate to its needs. It will have the opportunity to express its natural repertoire of behavior. A deficiency in any of these will result in stress, a physical response to poor welfare.

The consequences of stress can be extreme. If a wild male rat blunders into the territory of another male the dominant resident will attack it. The intruder will flee, and once it is clear of the territory it will be safe from further attack. But if we were to replicate this scenario in the laboratory, in an enclosure from which the intruder could not escape, we would see just how extreme the effect of unmanageable stress can be. In this situation the intruder rat's physical health would deteriorate very quickly and it may well die as a result of just a few hours of intermittent attacks, even though it had sustained no significantly damaging wounds. Fortunately the results of stress are rarely so severe this quickly. But even short-term stress can be sufficient to have a negative impact upon the immune system and a sufferer of prolonged stress will often exhibit physiological, physical and psychological damage. It is often the case that an individual will exhibit behavioral indicators of stress, useful cues to poor welfare, and in some cases an early warning system that can be used to mitigate a problem before it gets out of hand.

Animals that are confined for prolonged periods in suboptimal conditions may develop characteristically stereotyped patterns of behavior that they will repeat and repeat and repeat. These stereotypies might be as uncomplicated as repeatedly pacing a fixed route, or they may be quite elaborate involving a set routine of paces, head weaving, and other gestures. In some cases stereotypical rubbing, digging, biting, and chewing, etc. can even result in the performing animal damaging itself physically. Exactly what is going on in these situations is however the subject of some debate. Stereotypies will persist long after the circumstances that led to their development no longer apply, and so identifying where exactly poor welfare applies is often difficult. At a physiological level there is evidence that the performance of stereotypies actually reduces some of the symptoms of stress and so might be thought of as a coping strategy. (I'm sure that you have paced, or rocked, or chewed your nails, subconsciously, when waiting for important news or to face a difficult situation – all coping strategies to relieve stress.) This does not mean though that we can dismiss the poor welfare of an animal engaged in stereotypic behavior because it has solved the problem itself – it has not and we must not!

One approach that we might adopt in determining the optimal husbandry conditions for animals, and thereby ensure high welfare, might be to take a lead from the animals themselves. As a general rule of thumb when they are given a choice we should expect animals to choose what is good for them. They may even demonstrate that they are better able to make such choices themselves than we are to make them on their behalf. For example, in response to a UK government recommendation that battery cages for hens should have floors made of a heavy gauge rectangular mesh rather than a thinner hexagonal one because the latter would be uncomfortable, choice tests were carried out. The hens chose the hexagonal mesh, and it turned out that this flooring

Cheetahs (*Acinonyx jubatus*), in contrast to most members of the cat family, depend upon speed rather than stealth when hunting. Chases are generally short but they are clearly exhausting, because after a successful hunt a cheetah will rest for up to 30 minutes before eating (presumably only absolute exhaustion would delay feeding in an environment full of competitors ready to steal a kill). In captivity cheetahs are not given the opportunity to hunt live prey, to do so would be considered inhumane and would almost certainly be offensive to the viewing public. Quite apart from this, in countries such as the UK it is illegal. The lack of hunting opportunity in captivity results in lethargy, boredom, poor physical condition, and increased likelihood of stereotypies. In the case of the cheetah, a species whose liver is adapted for sudden mobilization of energy, lack of exercise may also be a major contributory factor in the development of liver disease. This is one of the main causes of death amongst captive cheetahs.

In an attempt to counter these problems the captive cheetahs at Edinburgh Zoo, Scotland, have been the subjects of a program of environmental enrichment. By providing the animals with a simulated hunting opportunity the research team involved hoped to increase the diversity of the behaviors exhibited by the animals (combating lethargy and boredom) and provide exercise opportunities. As an additional consequence, of course, the "improvement" in the behavior of the cats would increase their interest to the viewing public.

Prior to enrichment the animals were fed one rabbit per day and their food was thrown onto the floor in front of them. This particular cheetah enclosure is built on sloping ground. During the period of enrichment the mode of food delivery was modified. The rabbits were suspended from a wire that ran the length of the enclosure, parallel to the slope. Because the rabbit was introduced up-slope, gravity pulled it along the wire until it reached the fence at the down-slope end of the enclosure where a system of pulleys yanked it out of the cheetahs' reach. After a suitable period of training the cats learned to hunt and catch their "prey" before this happened.

During enrichment the animals would crouch some 5 meters from the point of rabbit delivery as soon as they saw the approaching keeper. Then they would wait until the rabbit had traveled some 10 meters along the wire before beginning their pursuit. Within 25-30 meters the cheetahs would draw level with their prey and fell it with a swipe of the forepaw, just as they would in the wild.

When the enrichment device was in use the cheetahs were seen to sprint three times as often as they had under their old feeding regime. So in terms of its aim to increase the exercise taken by the animals the program was a success. Enrichment did not increase the total range of behaviors exhibited by the animals, but it did change the relative proportions of time devoted to each of them and so it did increase the overall diversity of behaviors that were recorded. The way that the animals used the space available to them changed too. They used the various areas of their enclosure in a more diverse way and one that the research team felt represented improved space utilization. Perhaps most significantly the animals spent an increased amount of their time engaged in observation behavior following enrichment. In this context this meant that they spent more time on an elevated area of the enclosure staring outwards. In the wild cheetah use this technique to locate potential prey. In this study their gaze was directed almost exclusively towards an adjacent paddock stocked with blackbuck (*Antilope cervicapra*). Presumably the enrichment experience had re-kindled in them a motivation to hunt.

Williams, B.G., Warran, N.K., Carruthers, J. & Young, R.J. *et al.* (1996) The effect of a moving bait on the behavior of captive cheetahs (*Acinonyx jubatus*). *Animal Welfare*, **5**, 271–81.

provided their feet with better support than the "higher welfare" option that had been suggested.

Choice tests like these do, however, need to be carefully designed and evaluated if they are to prove useful. Care should be taken that the animals involved are aware of the choices available to them, and that they are "free" to exercise those choices. A great many species will for example avoid a novel situation and such behavior might compromise a test outcome. Hens given the choice between a familiar battery cage environment on the one hand and an unfamiliar outside run on the other will initially chose the familiar (against the predictions of most animal lovers). But if the test is repeated they will eventually modify the choice and take to the outdoor life.

Behavior and conservation

Traditionally conservation biology has been considered more the preserve of ecologists than of animal behaviorists. However, during the last decade a realization has developed that information about the behavior of animals, either as individuals or as part of a population of individuals, is invaluable in a wide range of conservation contexts.

As one of the main pressures contributing to increased rates of species extinction is habitat destruction and fragmentation it should come as no surprise that one of the main foci of conservation biology has been the design of nature reserves and protected areas. The theoretical background to the debate about reserve design has been grounded in island biogeography theory, and so the main questions asked have related to how big a reserve should be and how reserves should be arranged spatially. Is it better to have one big reserve or several smaller ones? Should reserves be isolated from one another, or connected via corridors to permit animal movements? Questions like these cannot be adequately answered without information about the dispersal patterns of individuals or about the likelihood that they will actually use the corridors as highways rather than perceiving them to be unsuitable "edge" habitat that they will not enter.

Another concern of conservation biology is the management of populations of animals, often with target population sizes in mind (relating to the carrying capacity of an area or to the minimum viable population of a species). In this context there is a tendency to consider all of the members of a population as being equivalent when in fact this is often not the case. Overexploitation of male animals in general may make it difficult for females to find and acquire mates. But the effect will be more marked in those species where females exercise mate choice based

Key reference
Caro, T. (1999) The behavior–conservation interface. *Trends in Ecology and Evolution*, **14**, 366–9.

Link
Not all of the members of a sex are equally attractive as potential mates.
Chapter 8

on ornamentation (horns, etc.) that also determine a hunter's choice of target.

I cannot think of a better example of the need to be fully aware of the reproductive behavior and mating system of an exploited animal than the case of the lion. In a number of reserves the licensed hunting of "trophy" lions (males) represents an important income stream that contributes to the continued existence of the reserve. But shooting males necessarily increases the rate at which prides of lions are taken over by new males. New males commit infanticide to ensure that they do not raise the young of their rivals and to ensure that the pride females will come into estrus as soon as possible. This of course means that the impact upon the population of shooting one male is larger than the removal of that male. It is likely that this problem could be further exacerbated if males are repeatedly shot in the same area, perhaps because it offers ease of access to the hunter.

As a strategy of last resort managed populations are often augmented by the reintroduction into the wild of stock raised in captivity. During the captive breeding component of these programs failures may be the result of inadequate opportunities for effective courtship, particularly if male–male competition is important and too few animals are available. When young are produced the captive situation often results in a failure on the part of the parent to raise them. In these situations animals may be raised by a surrogate species, or by humans. This leads to the risk that they will imprint upon an inappropriate model "parent" and as adults will exhibit inappropriate reproductive behaviors. For example as a result of experiments that involved using sandhill cranes (*Grus canadensis tabida*) to foster whooping cranes (*Grus americanus*), a generation of cranes that had difficulty forming relationships with their own kind was produced. As a conservation "product" these animal were clearly of limited value. In the case of this particular species the problems that imprinting can cause have been successfully overcome by the International Crane Foundation in Wisconsin. Young chicks are reared in isolation using glove puppets that mimic the heads of their parents. The chicks are then moved to a suitable habitat where they are exposed to model cranes and to a human attendant wearing a crane costume. This person guides the birds to suitable feeding areas whilst playing tape recordings of crane contact calls. Occasionally the birds will also be played a crane alarm call to coincide with the appearance of an uncostumed human just to make sure that they will learn

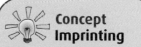

Concept Imprinting

As a concept imprinting was first described in birds. It refers to a particular form of learning that occurs during a usually very short **sensitive period** early in the life of an individual. At this time critical information like species identity, the identity of one's parent, the identity of suitable reproductive partners, and in some cases the characteristics of an appropriate habitat in which to live, are learned.

to avoid people in future. This is an extremely labour-intensive procedure, but given that the result has been birds able to survive in the wild, to join flocks of wild birds, and to learn appropriate migration routes, it does work.

Summary

In this chapter behavior has been defined and a broad overview to the methodologies that we use when we study it has been provided. Specifically the idea that animal behavior is a robust science has been reinforced through a demonstration that it follows the tried and tested scientific method of observation leading to hypothesis generation and experimental testing. It has been shown that it is essential that we consider behaviors as being adaptations, the products of evolution by natural selection. In addition, the study of animal behavior has been demonstrated to have a clear relevance to several areas of our daily lives.

Questions for discussion

As a group imagine that you have been moved from your terrestrial environment to a life in the deepest ocean. In what ways would you need to modify your behavior to ensure your survival?

Do you think that squid ink because they are afraid? Remember that fear is a human emotion and that this is an anthropomorphic question.

Which of the following is the best way to study behavior? By direct observation, in an experiment, by a cost–benefit analysis, or by the comparative approach?

2 Controlling Behavior: The Role of the Nervous System

Certain actions . . . are the direct result of the constitution of the nervous system, and have been from the first independent of the will.

Charles Darwin, 1872

It should be obvious from the previous chapter that the behavior of an animal is a complex affair and one that requires a high degree of control and coordination. Of course it must also require an amount of information about both the individual's surroundings and its internal state. Although some hormone-mediated short-term control does take place (this will be discussed in Chapter 3), the proximate (short-term) control of behavior is largely the job of the specialist cells of the nervous system.

Contents

Key points

• At a proximate level the behaviors animals perform are controlled in part by the activity of the nervous system. This network of specialized cells provides a means of rapid information transfer within the animal, linking sensory input to central processing and motor responses.

• Many types of behavior can be described as reflexes, relatively simple involuntary responses to stimuli. In some cases these stimuli (releasers) are very specific, in other cases they are more general.

• Animals possess a range of strategies to maximize the efficiency of their information gathering, and their behavioral performance.

• It is important to remember that nerves alone do not control behavior, the interaction between animal and environment is vitally important.

Stimulating a behavior

Fixed action patterns

It is entirely possible that I may be able to cause you to perform a very specific and predictable pattern of behavior right now as you read this book. Look at the face in Plate 2.1. What is it doing? It is performing a yawn. Yawns are members of a class of behaviors termed fixed action patterns (FAP) by ethologists. An FAP is an **instinctive** behavior. It is performed perfectly first time, without practice and without any tuition (as we will see in Chapter 4 contrary to popular belief instinctive behaviors are not "genetically determined" and they can be modified).

A yawn lasts for about 6 seconds and involves a fully open mouth and in many cases the closing of the eyes. Once started it is difficult if not impossible to stop. So FAPs can be described as always running to completion. However FAPs, and therefore yawns, have one other important characteristic – something triggers their performance. This trigger or stimulus is usually referred to as a releasing mechanism or a releaser. So now go back to the photograph of the yawn. The thing about yawning is that it is contagious. A yawn is a releaser for another yawn. In fact it is often the case that hearing a yawn or just seeing a photograph of a yawn is enough to release the behavior. So did it work yet? Have you yawned? If you did look at the people around you, are they yawning too?

Young herring gulls (*Larus argentatus*) hatch from their eggs able to do little more than sit up. They are completely dependent upon their parents for food and protection for some weeks. Adult gulls feed their young by regurgitating food directly into their beaks and the regurgitation reflex is initiated in response to a releaser provided by the chick when it pecks at the tip of the adult bird's beak. But what stimulates the pecking reflex in the chick in the first place? Niko Tinbergen found that young gulls do not need to see the whole parent to begin pecking, in fact a model of the head and beak of the bird is a perfectly adequate stimulus. Taking advantage of this fact he went on to carry out a number of experimental studies that enabled him to shed further light on some of the general properties of releasing mechanisms.

An adult herring gull's beak is bright yellow with a red spot close to its tip (Plate 2.2). Tinbergen manipulated this very distinctive pattern in his models and found that the presence of the

Plate 2.1 Yawning is contagious! © G. Scott.

Plate 2.2 The red spot on the beak of an adult herring gull is an irresistible target for a chick. © G. Scott.

spot was very important. The chicks were far more likely to peck at beaks with spots than those without. They were also more likely to peck at spots that contrasted strongly with the beak color, and interestingly at red beaks rather than yellow ones. In fact a pencil with some red bands on it moved horizontally in front of the bird will elicit the biggest response of all. Based upon these results we can deduce that the chick must: (i) have an innate (inborn) ability to recognize appropriate stimuli; (ii) that the behavior is an innate reflex released by an appropriate stimulus; (iii) that the chick doesn't need all of the other cues that describe a parent, and that it only focuses upon the stimulus itself; and (iv) because non-herring-gull-type stimuli did the job as well if not better, that there must be a degree of flexibility in the system. This is an important point and one that we will return to in Chapter 4.

So we can see that a stimulus releases a behavior. But how? Presumably the herring gull chick sees the spot on the beak of the parent bird, and we know that its response is to peck. But what takes place within the animal that completes this sequence of events? In this particular case we just don't know, although we can be sure that the nervous system is involved and as we will see in the following sections of this chapter there are some behaviors that we do understand in far greater detail at this level.

Link
The gull chicks' pecking reflex is not as invariant as might at first be believed.
Chapter 4

Components of the nervous system

Almost all discussions of the nervous control of behavior start in the same place, with a small hammer tapping the human leg just

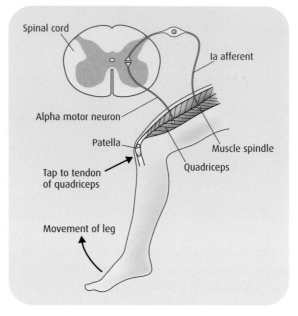

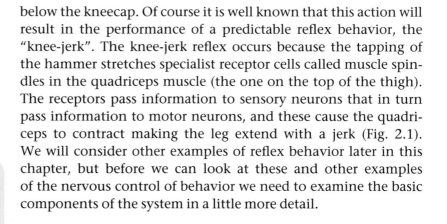

Fig. 2.1 The human knee-jerk reflex. (From Bear et al (2001) *Neuroscience: Exploring the Brain.* 2nd edn. Lippincott, Williams & Wilkins, Baltimore.)

below the kneecap. Of course it is well known that this action will result in the performance of a predictable reflex behavior, the "knee-jerk". The knee-jerk reflex occurs because the tapping of the hammer stretches specialist receptor cells called muscle spindles in the quadriceps muscle (the one on the top of the thigh). The receptors pass information to sensory neurons that in turn pass information to motor neurons, and these cause the quadriceps to contract making the leg extend with a jerk (Fig. 2.1). We will consider other examples of reflex behavior later in this chapter, but before we can look at these and other examples of the nervous control of behavior we need to examine the basic components of the system in a little more detail.

Neurons

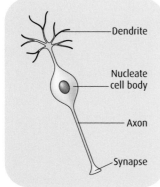

Fig. 2.2 The gross morphology of a neuron.

Neurons are the main building blocks of the nervous system. These are the cells that receive, sort and pass on the information that in approximate sense cause an animal to behave. Neurons are nucleate cells like any other (Fig. 2.2), but they possess specialized processes that enable them to acquire information (the dendrites) and to transmit it often considerable distances through the body (usually via the axon). For example, the medial giant interneurons of the crayfish are in the animal's brain, but their axons extend all of the way to the tip of its tail. At another extreme, however, are the amacrine cells, neurons that are only involved in local signaling and in which the axon is much reduced if it is present at all. The processes of these cells are all dendritic.

Three main classes of neuron are involved in the performance of behavior. Sensory neurons collect information from specialized receptor cells in the nervous system. Motor neurons connect directly to muscles and stimulate them into action. Interneurons form links between these inputs and outputs.

The resting potential

The information handled by neurons takes the form of electrical signals. These are propagated as small changes in voltage between the inside and outside of the cell. To facilitate this the cell membranes of neurons are permeable to ion flow. Figure 2.3 shows a simplified representation of a section of neuron cell wall. Note that there are sodium ions on both sides of the membrane and that the membrane is permeable to those ions (it has "gates").

Two different gradients of ion movement across this membrane exist. On the one hand, ions are moving out of the cell down a

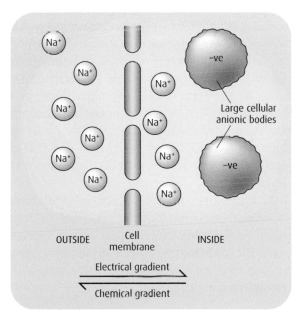

Fig. 2.3 The chemical and electrical gradients relating to sodium ions (Na^+) across the cell membrane of a hypothetical nerve cell. (From Carew, T.J. (2000) *Behavioral Neurobiology*. Sinauer Associates, Sunderland, MA. Reproduced with permission of Sinauer Associates.)

chemical gradient that would, in the absence of any other force, eventually result in a similar concentration of the ions on both sides of the membrane (i.e. diffusion is taking place). On the other hand, there is also an electrical gradient within the cell as a result of the difference in charge between the sodium cations and larger anionic units (such as cellular proteins) that are too big to pass out of the cell. As a result of these two opposing forces ions flow in both directions across the membrane, eventually achieving equilibrium. We term this the equilibrium potential.

When we consider the movement of ions into and out of cells in this context we really only need to consider sodium (Na^+), potassium (K^+), chloride (Cl^-), and calcium (Ca^{2+}). Each of these ions will of course achieve its own equilibrium potential and summed together they amount to the **resting potential** of the cell. This can be thought of as the "force" required to maintain the status quo and it is usually measured as having a value of around −65 millivolts. This is the cell's background state if you like, and if the cell is involved in electrical signal transmission, as either a sender or a receiver, we will be able to record a deviation from that state.

The action potential

If the interior of a neuron is made to have a greater positive charge than the resting potential the membrane will depolarize. Similarly the application of an increased negative charge within the cell will result in membrane hyperpolarization. Depolarizations tend to be excitatory in nature, stimulating activity in the neuron, whereas hyperpolarizations tend to be inhibitory. If the charge change related to a depolarization is sufficiently large, a very particular electrical phenomenon called a **spike** will be triggered. As Fig. 2.4 illustrates spikes, or **action potentials** as they are more properly termed, follow a characteristic pattern of rising and falling voltage. This is an all or nothing phenomenon that is always triggered when the action potential threshold is reached; it always follows the same waveform.

Sometimes a single postsynaptic potential (PSP) is enough to induce a spike, but often the combined effect of a number of them is required. Some nerve cells exhibit a phenomenon termed **summation** whereby a series of small PSPs coming from one or a number of sources have an additive effect. Their individual charges are summed to achieve the threshold required for spike production.

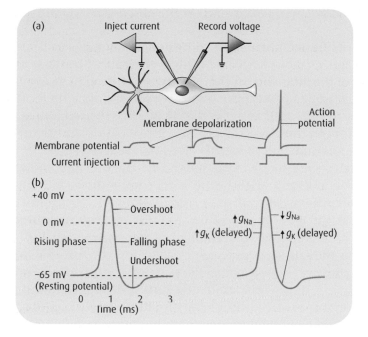

Fig. 2.4 The action potential. (a) When a small pulse of positive current is injected into a neuron, small depolarizations are produced. Once threshold is achieved, an action potential is generated. (b) The different phases of the action potential (top) and the ionic events underlying it (bottom). (From Carew, T.J. (2000) *Behavioral Neurobiology*. Sinauer Associates, Sunderland, MA.)

Other cells require a series of PSPs that rather than being added together each elicit a slightly larger PSP than themselves until the action potential threshold is reached. This phenomenon is termed **synaptic facilitation**.

The spike lasts for just one or two milliseconds, but can move along an axon at speeds in excess of 100 m/s. It begins as a rapid increase in voltage through a rising phase to an overshoot phase when the interior of the cell is more positively charged than the exterior. In terms of ion transport once the depolarization of the cell membrane reaches the action potential threshold gates in the membrane open, allowing Na^+ to enter the cell, increasing the measured voltage very rapidly. The increasing Na^+ conductance is followed by an increase in K^+ conductance out of the cell as an attempt is made to return to equilibrium. The Na^+ gates now begin to close (termed Na^+ inactivation) and the measured voltage begins to fall. It falls below the resting potential as a result of the ongoing activity of the K^+, but as the depolarization of the membrane subsides so too does the conductance of K^+ and so the resting potential is re-established.

The synapse

Neurons transmit messages to one another at specialized junctions termed synapses. The cell delivering the message is termed the presynaptic cell and the one receiving it is the postsynaptic cell. The area between the two is termed the synaptic cleft. The specialized areas of cell membrane at the synaptic junction are therefore referred to as the pre- and postsynaptic membranes.

The transmission of signals across some synapses is electrical. In such cases, termed gap junctions, there is a physical link between one cell and the next to permit direct current flow between them. This allows particularly rapid cell to cell signal transfer and is a characteristic of the systems requiring the simultaneous coordination of a number of components, such as is the case in the tail-flip escape response of the crayfish that we will consider later in this chapter. The majority of synapses however rely upon a slower (but still pretty fast) system of chemical transmission that involves the release into the synaptic cleft of specialist neurotransmitters (which may be amines, amino acids, or peptides).

When an action potential reaches the synaptic area of the signaling cell depolarization of the presynaptic membrane will occur. This change in the voltage across the membrane causes channels, or gates, to open that are permeable to Ca^{2+}. This permits a net flow of Ca^{2+} into the cell in an effort to re-establish the resting potential. The increased Ca^{2+} in the cell triggers a chain of events that culminates in the secretion of the neurotransmitter into the synaptic cleft.

Neurotransmitters diffuse quickly across the synaptic cleft and dock with receptors on the postsynaptic membrane. Here too channels open and ion flow occurs. Depending upon the particular ions that are involved, this results in either the depolarization or hyperpolarization of the postsynaptic cell. Depolarizations will result in excitatory postsynaptic potentials (EPSPs), which make it more likely that a spike will be triggered and the signal will be passed on down the line. Hyperpolarizations result in inhibitory postsynaptic potentials (IPSPs), which reduce the likelihood of spike production. So exactly how likely it is that a spike in a presynaptic cell will stimulate a spike in the postsynaptic cell with which it communicates will depend upon the precise combination of EPSPs and IPSPs involved – a process termed **integration**.

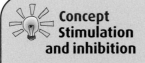

**Concept
Stimulation
and inhibition**

When the membrane of a neuron depolarizes the cell is stimulated and signal transmission will take place. However some neurons inhibit signal transmission in their target cells. This happens because their neurotransmitters hyperpolarize the membrane.

Application **Neurobiology and behavior**

The human brain contains approximately 10^{12} nerve cells and so it might seem that trying to understand the neural machinery of our own behavior is a task that is beyond our abilities. Fortunately nerve cells are all remarkably similar to one another in terms of their mode of operation across the whole of the animal kingdom.

This means that the neurobiological and technical breakthroughs that are currently being made in this exciting and fast moving area of animal behavior could have a real impact upon our understanding of our own machinery for behavior.

Controlling prey capture

The ability to correctly identify, acquire and process food is a behavioral process that is surely a prime candidate for reflex status. Many animals, and particularly those that never meet their parents, need to be able to forage for themselves at the very earliest opportunity. If they cannot they may well starve. Foraging behaviors in general will be considered in detail in Chapter 6, but at this stage it will be useful to consider some examples of their neurological control.

Case study "Touch" at a distance: the trichobothria of the wandering spider

The wandering spider *Cupiennius salei* captures its prey without the aid of a web. Although the majority of the invertebrates that it captures in the wild are those that walk, it is also able to capture flying insects. It does this with a spectacular and very precise leap into the air. Given that spiders have eight well-developed eyes, we might assume that they use visual cues to carry out this remarkable feat. However Freidrich Barth and his colleagues have demonstrated that this is not the case. In an experiment they covered the eyes of a hungry spider to temporarily blind it. Despite this incapacity the spider was able to detect airborne prey animals up to 20 centimeters away and to capture them with a leap of several centimeters. By contrast, web-based spiders are able to locate prey a mere 1 centimeter away. Of course this difference between the two spider types is not surprising, web-based spiders do not need to detect moving prey because they rely upon their web. On the other hand, the advantages of long-distance prey detection to a mobile hunter are obvious. There is also a basic physical difference between the two spider types. Wandering spiders are far more hairy than their web-based cousins. In fact *Cupiennius* is one the most hairy spiders of them all. The hairs in question are termed trichobothria and *Cupiennius* has close to 1000 of them arranged in groups of 2–30 on their pedipalps and walking legs. If all of the trichobothria of a spider are removed it loses the ability to locate or respond to prey stimuli. If the trichobothria of only one side of the body are removed the spider will respond to the presence of prey, but its ability to locate

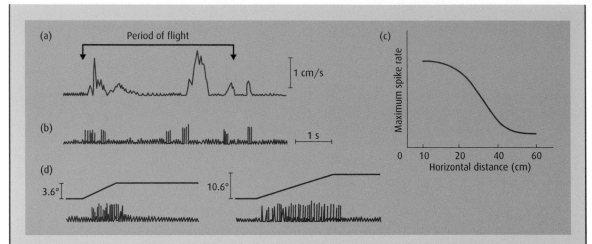

Fig. 2.5 The response of a single trichobothrium to the wind generated by a tethered fly. (a) Wind speeds recorded from the fly. (b) Action potentials recorded from the trichobothria. (c) The maximum spike rate of a trichobothrium in response to winds produced by a fly at various distances from the spider. (d) The response of a trichobothrium to linear displacements of 3.6° and 10.6° (top lines), the bottom lines show spike production. (From Barth & Höller (1999). Reproduced with permission of the Royal Society.)

the source of the stimulus is impaired and it will always orientate towards its intact side irrespective of the position of the prey. Cleary these hairs play a key role in prey capture. Filiform hairs are commonly used as sense organs to detect motion. In the insects each hair is supplied by a single sensory cell that responds to both the strength and the direction of the stimulus. In spiders, however, the arrangement is slightly different with each hair supplied by four cells each of which responds when the hair is deflected in a particular direction. A flying insect generates behind and below it a "cone" of wind (broader closer to the fly and tapering away from it) and so it would seem likely that the trichobothria are involved in some assessment of this air flow (Fig. 2.5).

Experiments involving a tethered spider allowed Barth and his team to monitor the pattern of excitation of the sensory cells associated with a single trichobothrium when a fly attached to a wire was placed above it. They found that it responded most strongly when it was close to the center of the cone of wind and less strongly if it was moved towards the edge of the cone, or out of it altogether. They also recorded a stronger response to closer flies than to more distant ones, and from winds with higher frequencies and velocities.

How then does the spider use this information to identify and target the fly? Based on their experimental observations of the functioning of individual trichobothria and their extensive knowledge of the behavior of *Cupiennius*, Barth and his colleagues suggest that:
- the trichobothria allow the spider to distinguish between the relatively high frequency wind produced by a flying insect (up to 150 MHz) and typically lower frequency background air movements (about 10 MHz) characteristic of the spider's environment, and;
- that the spider is able to simultaneously compare the individual excitation patterns of its many trichobothria to exactly target its prey.

Barth, F.G. & Höller, A. (1999) Dynamics of arthropod filiform hairs. V. The response of spider trichobothria to natural stimuli. *Philosophical Transactions of the Royal Society of London. Series B: Biological Sciences*, **345**, 183–92.

Barth, F.G., Wastl, U., Humphreys, J.A.C. & Devakonda, R. (1993) Dynamics of arthropod filiform hairs. II. Mechanical properties of spider trichobothria (*Cupiennius salei* Keys.). *Philosophical Transactions of the Royal Society of London. Series B: Biological Sciences*, **340**, 445–61.

Prey capture by toads

There are a great many reflex behaviors that one could describe, and in many cases they may have been analyzed to the point that we can say something about the releasing mechanisms and broader neural systems that are involved. Jörg-Peter Ewart and his coworkers have gone one step further. The thrust of their work in this area has been to link the behavioral aspects of the toad prey capture reflex to a particular area of the central nervous system, its **feature detector.**

When a prey item (usually a worm or an insect) moves through the visual field of a toad a sequence of behaviors will be performed:
1. The toad moves so that it faces the prey animal.
2. The toad approaches the prey animal to within striking distance.
3. The toad uses its tongue to strike at the prey animal and captures it.
4. The toad swallows the animal.
5. The toad wipes its mouth.

This behavioral sequence can be regarded as a reflex because it is released by the visual stimulus of the prey. It is innate because a newly metamorphosed toad that is completely prey-naïve will perform the whole sequence. Although in one respect the behaviors are very rigidly performed (if the prey disappears during the first stage of the sequence the toad will still run through a "capture" even to the point of wiping its mouth), there is an element of flexibility in terms of the start point. If the prey animal is directly in front of the toad and within striking distance, it misses out steps 1 and 2.

If a hungry toad is presented with a model "worm" (nothing more complicated than a strip of card that contrasts with its background and is longer than it is tall) the reflex will be triggered. By taking advantage of this it is possible to present a range of worm-like and nonworm-like stimuli to better understand the behavior of the toad. Ewart did this and arrived at the results presented in Fig. 2.6, from which he deduced that the more worm-like a stimulus is the stronger its releasing effect. The discrimination exhibited suggests that a stimulus filter and feature detector may be at work here.

The toad's response to the square stimulus is interesting. Presumably small squares resemble small beetles or other insects and are therefore fair game, but above a certain size the square

Concept
Stimulus filtering and feature detectors

An animal's senses are tuned to its way of life. The neurons of the sensory system act as hierarchies of stimulus filters, "deciding" which components of the wealth of information they receive to consider and which to ignore. In this way information specific to a particular behavior is channeled along a specific neural pathway. Those areas of the nervous system that respond to particular stimuli are termed feature detectors.

Key reference
Ewart, J-P. (1997) Neural correlates of a key stimulus and releasing mechanism: a case study and two concepts. *Trends in Neuroscience*, 20, 332–39.

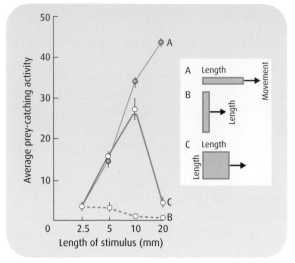

Fig. 2.6 The responses of a toad to moving stimuli of gradually increasing size. A, a worm-like stimulus; B, a non-worm-like stimulus; C, a square stimulus. No response was recorded to any stationary stimuli. (From Ewart, J-P. (1985) Concepts in vertebrate neuroethology. *Animal Behaviour*, **33**, 1–29. Reproduced with permission of Elsevier.)

represents too much of a mouthful for even a predator, and so turning and running away might be the more appropriate response.

There is no doubt that the eyes are the important sense organs in the case of this behavior, and so the optical system of the toad is the obvious place to search for the stimulus filter and the feature detector. Information enters the system as light via the eye. It is translated into a pattern of electrical signals via the rod and cone cells of the retina. They in turn stimulate the cells forming the retinal ganglion and the signal is transmitted onwards to the brain via the axons that comprise the optic nerve. These axons connect to two regions of the brain, the thalamic pretectum and, behind it, the tectum. The orderly way in which these connections are made results in a pair of "maps" of the retina on each region of the brain. Arrangements like this, with neurons originating in close proximity to one another terminating in close proximity, are not uncommon in systems of this type. The tectum itself is composed of layers of neurons, and axons from six different classes of retinal cell terminate in specific layers (Fig. 2.7).

The stimulus filter

Information received by the toad retina is transmitted to the optic tectum of the brain via the optic nerve, a bundle of axons originating from a class of cells termed optic ganglia. Information is

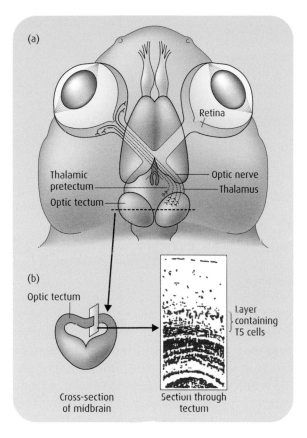

(a)

Retina

Thalamic pretectum

Optic tectum

Optic nerve

Thalamus

(b)

Optic tectum

Layer containing T5 cells

Cross-section of midbrain

Section through tectum

Fig. 2.7 (a) The toad brain showing the connections between the retina and the thalamic pretectum and optic tectum via the optic nerve. (b) A section taken through the brain to reveal the position of the T5 cells. ((a) After Camhi, J.M. (1984) *Neuroethology: Nerve Cells and the Natural Behaviour of Animals*, p. 231, Sinauer Associates, Sunderland, MA; data from Muntz, W.R.A. (1964) Vision in frogs. *Scientific American* **210**(3), 110–19. (b) After Camhi, J.M. (1984) *Neuroethology: Nerve Cells and the Natural Behaviour of Animals*, p. 231, Sinauer Associates, Sunderland, MA; data from Ewart, J.P. (1980) Concepts in vertebrate neuroethology. *Animal Behaviour*, **33**, 1–29.)

received by these ganglia from bipolar cells (interneurons) that are in turn stimulated by the receptor cells (rods and cones) of the retinal surface. As Fig. 2.8 shows each ganglion receives its information from a discrete area of the retinal surface referred to as a **receptive field**.

The receptive field of each ganglion cell comprises an area of the retina divided into two concentric circles (Fig. 2.8). If the whole of the receptive field is illuminated, all of the retinal receptor cells will transmit a signal to the ganglion via the bipolar cells. This will stimulate the ganglion cell and an electric potential is recorded in its axon, and information is transmitted to the brain. However, a far stronger signal will be transmitted to the brain if only the cells of the inner circle of the receptive field are stimulated. Why then is the signal diminished when both circles are stimulated? The answer to this question can be found if only those cells comprising the outer circle are stimulated. The effect

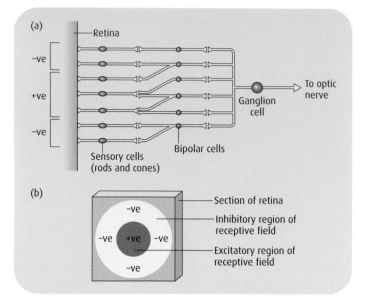

Fig. 2.8 The excitatory and inhibitory regions of the retinal receptive field of a single ganglion cell. (a) The arrangement of cells involved in the transmission of sensory information from the retina to the optic nerve. (b) The arrangement of the excitatory and inhibitory areas on a retinal section. In both figures –ve and +ve refer to inhibitory and excitatory areas respectively.

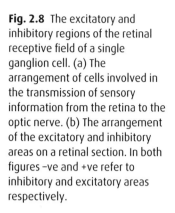

 **Concept**
Receptive field

The receptive field of a cell can be defined in two ways. It is that area of the world of the animal from which information is received by that particular cell. It is also the specific region (or group of sensory cells) of the information receptor (the retina for example) from which information is transmitted to the cell.

of this is that no signal is transmitted to the brain because the ganglion cell is not stimulated. The inner circle is an excitatory region of the retina, whilst the outer circle is an inhibitory region. If both are stimulated they cancel one another out (but not completely in this case because some stimulation did still occur).

In terms of what the animal sees we can comprehend the consequences of this arrangement of differential excitation and inhibition if we think about two objects of differing size passing through the visual field of a toad. If a small object moves in front of the toad light reflected from it on to the retina will focus on a small area of the retinal surface. This means that there is a good chance that a greater number of the excitatory retinal cells will be stimulated, these ganglia will transmit strong signals, and the toad will see a potential prey item. On the other hand, a very large object passing in front of the toad will stimulate a larger area of the retina and will therefore stimulate both excitatory and inhibitory receptors and either a weak signal or no signal will reach the brain. This arrangement will therefore allow the toad to discriminate between potential prey items (small objects) and nonprey items (larger objects). If it can discriminate between these two classes of stimulus it can behave appropriately towards each of them. So this arrangement of retinal cells acts as a highly effective stimulus filter.

The feature detector

Each retinal cell class (referred to as R1 to R6) exhibits a particular response to a specific kind of stimulation. By monitoring the activity of the cells in the toads involved in his experiments, Ewart found that R2, R3 and R4 cells were the most active and therefore the most likely to be involved in prey detection. Further work, however, showed that none of them could be a feature detector, because although they did all respond in predictable ways to the worm and nonworm stimuli, no one class showed a "preference" for one particular stimulus. Ewart therefore concluded that although the retinal cells were responding to information about contrast, movement, and stimulus size, the feature detector must lie within the brain itself.

The first area to fall under investigation was the thalamic pretectum. Remember that this area of the brain has the retina "mapped" onto it, and throughout this map Ewart found a class of cells (termed TP3) that are responsive to visual information. However, analysis showed that these cells too could not be the feature detector because their pattern of activity did not match the behavioral observations made earlier (Fig. 2.9). Moving further into the visual system attention eventually fell upon the tectum itself, and this time upon a class of cell termed T5(1) and

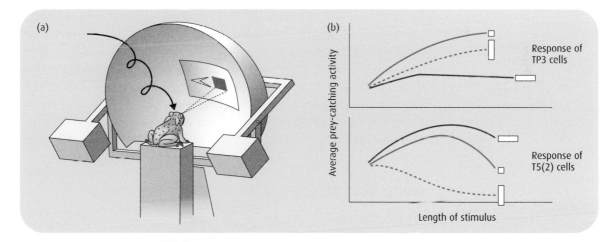

Fig. 2.9 (a) Measuring the response of TP3 and T5(2) cells. (b) Comparison of the behavioral response of a toad to varying moving stimuli with the activity of the animal's TP3 and T5(2) cells. Refer back to Fig. 2.6 to compare these patterns to the behavior of the toad. (From Ewart J-P. & von Wietershein, A. (1974) Pattern analysis by tectal and thalamus/pretectal nerve nets in the visual system of the toad *Bufo bufo* (L.). *Journal of Comparative Physiology*, **92**, 131–48. Reproduced with permission of Springer-Verlag.)

T5(2) found deep within its layers. Observations of the firing patterns of the latter produced results suggesting that they were indeed the feature detector (Fig. 2.9), responding most strongly to worm-like and least strongly to nonworm-like stimuli.

Now that the crucial cells have been identified we can pay more attention to their activity in terms of their interaction with other areas of the brain. Specifically it turns out that Ewart was not wrong to suspect the involvement of the TP3 and T5(1) neurons in prey detection and capture. As a result of experiments involving firstly the isolation of various components of the system and then their artificial stimulation via electrodes, it can be shown that TP3 cells have an inhibitory effect and T5(1) cells have an excitatory effect upon T5(2) cells.

In response to a worm-like stimulus the T5(1) cells are able to stimulate the T5(2) cells without conflicting inhibition from the TP3 cells, and so the net effect is a strong T5(2) response and the facilitation of the prey capture reflex. However when a nonworm-like stimulus is presented the T5(1) cells respond weakly and do not counter the strong inhibitory effect of the TP3 cells. Therefore the T5(1) cells do not respond and the stimulus is ignored. The various sizes of square stimulus elicit different results depending upon the balance of T5(1) excitation and TP3 inhibition of the T5(2) cells.

So by a series of whole-animal observations and neurological investigations it has been possible to link observed behaviors to particular groups of cells and to demonstrate the existence of both a stimulus filter and a feature detector. This system is far less simple than one might have at first assumed. It is not just a case of one cell "recognizing" a worm and another a square and so on, but rather a group of cells working together to filter out nonuseful information and then to analyze the remainder in a way that in this case results in the recognition of potential food.

Controlling escape behavior

Command neurons

The term "command neuron" has been used to describe a single neuron or a small set of connected neurons that elicit specific behavior in response to a natural stimulus. The relative simplicity that a pattern of organization such as this appears to possess is

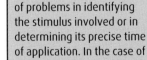

Concept Latency

The latency of a behavior refers to the time lag between the application of a stimulus and the observation of the response. It is often difficult to measure the latency of a complex behavior because of problems in identifying the stimulus involved or in determining its precise time of application. In the case of reflex behaviors it is usually the case that the stronger the stimulus, the shorter the latency.

Case study Escape jetting: can a cold squid still flee?

If I were to clap my hands together sharply with an untied, water-filled balloon between my palms a pressurized jet of water would shoot out. The same simple principle of physics is exploited by the squid *Loligo opalescens* when it is startled.

In response to the startling stimulus a single spike is produced in the squid's giant axon system. This spike innervates the circular muscles of the animal's mantle, causing it to undergo a strong and sudden twitch, rapidly increasing the internal hydrostatic pressure, and forces a jet of water out of the animal's funnel. This jet of water propels the squid to safety at some considerable speed. This apparently simple reflex response has the key features of a good escape behavior, short latency, and stereotypic simplicity. However, because of the environmental conditions experienced by this animal the performance of this particular behavior may not be as straightforward as one might at first assume.

Loligo opalescens is a pelagic animal. This means that it lives in the open sea, and in this case moves freely between the waters of the surface and the deep (to depths of several hundred meters). Therefore during the course of a single dive an individual might encounter a gradient of water temperature from 14°C at the surface to as little as 6°C. Generally the activity of animal systems, and this would include the activity of nerves and muscles, is strongly temperature dependent. So it would seem reasonable to assume that *Loligo* must posses some physiological and/or behavioral adaptation if it is to jet to safety in deep and shallow water with equal efficiency.

Through a series of experiments involving live animals and *in vitro* preparations of muscles and their nerves, a team of scientists based at the John Hopkins Marine Station in California have sought out these adaptations. They have observed that at lower temperatures the latency of jetting behavior increases, probably due to the direct effect of temperature on processes such as the transmission speed of signals through nerves and the speed at which muscles contract. But despite their slower reactions the squid in colder water still managed to "escape" as well as their warm-water counterparts. This was because their maximum swimming velocity was higher, and the distance they covered was greater. So the cold squid may be slow to get going – but they go further, faster.

As I write the exact mechanism at work here has yet to be confirmed, but the evidence presented by the John Hopkins team strongly supports their idea that it involves a coordinated alteration in the recruitment of the nerves involved in the behavior. At higher temperatures the giant axon alone seems to be involved, producing a single muscle twitch and a single sharp increase in water pressure within the mantle cavity. At lower temperatures the slower activity of the giant axon seems to be augmented by secondary nongiant axon nerve activity and an associated secondary, prolonged, increase in pressure. It seems likely that this can happen because inhibition of the nongiant axon nerve activity by some component of the central nervous system is deactivated as a response to low temperature.

Neumeister, H., Ripley, B., Preuss, T. & Gilly, W.F. (1999) Effects of temperature on escape jetting in the squid *Loligo opalescens*. *Journal of Experimental Biology*, **203**, 547–57.

thought to allow an efficiency and speed of response to a stimulus that may not be possible from a less directly controlled system. Although it is now recognized that a wide range of behaviors are controlled via command neurons, which are in turn organized in a variety of ways, the most thoroughly described system is that which elicits the tail-flip escape response of the crayfish. Historically this behavior has some importance too, being the first to be described using the term "command neuron" when it was coined in 1964 by C.A.G. Wiersma.

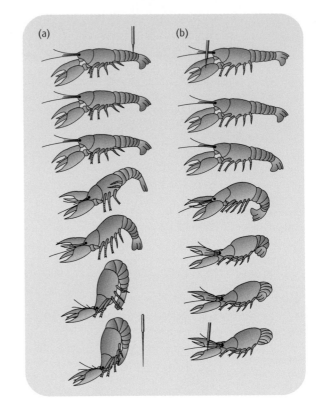

Fig. 2.10 The rapid tail-flip escape response of crayfish. Each sequence can be completed in as little as 70 ms. Sequence (a) is the lateral giant interneuron-mediated response resulting from tactile stimulation towards the rear of the animal and sequence (b) is the medial giant interneuron-mediated response to stimulation towards the front of the animal. (From Wine, J. & Kranse, F.B. (1972) The organisation of escape behaviour in the crayfish. *Journal of Experimental Behaviour*, **56**, 1–18. Reproduced with permission of The Company of Biologists Ltd.)

Flipping crayfish: a rapid escape response

In the same way that the ability to recognize and respond to the proximity of potential prey is important to a hungry animal, it is also vitally important that it is able to recognize and escape from potential predators. Figure 2.10 illustrates the rapid escape behavior, the tail-flip, of the crayfish. This behavior is a highly stereotyped response to a mechanical stimulus such as the beak of a hunting heron. Flipping the highly muscular tail propels the animal away from the source of danger, and as one would perhaps expect the onset of the behavior following stimulation is extremely rapid. It has a latency of as little as 10 milliseconds. A very similar sequence of behavior can be elicited as a result of the direct electrical stimulation of the animal's nervous system.

The tail-flip is produced by the rapid contraction and re-extension of the massive muscles that fill much of the abdomen of the animal. (This of course is the bit of the crustacean that the predator is after, and the bit that many humans enjoy too.) The activity of these muscles is mediated by a number of large motor neurons on each side of each segment of the abdomen. In addition to these, one exceptionally large neuron called the **motor giant** sends an axon branch to every one of the muscle fibers involved (it has a parallel inhibitory motor neuron). These and a number of less important motor neurons are excited via giant interneurons (GIs).

Crayfish possess two pairs of GIs that run the entire length of the animal's nervous system, from its head to the tip of its tail. In addition to their being considerably larger than other nerve cells, these GIs enable a signal to travel through the body very quickly. One of the pairs, the medial giant interneurons (MGIs), have their cell bodies and dendrites in the brain of the animal, and their axons extend all the way to the tail. The other pair, the lateral giant interneurons (LGIs), have a very different structure. They are a chain of individual cells, one in every segment of the animal and linked to one another by synapses. In addition to their having different structures, the two GI systems have different connectivities (Fig. 2.11) and slightly different functions. The activity

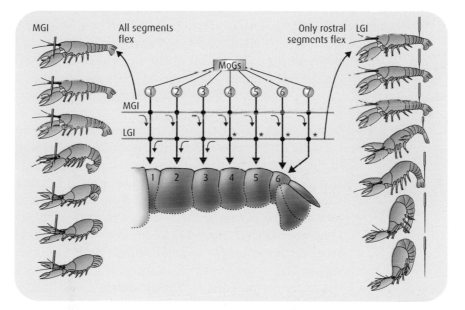

Fig. 2.11 The MGIs making synaptic contact with all of the motor giants (MoGs; solid circles), with the result that during an MGI tail-flip all six of the abdominal segments flex and the animal is propeled backwards. LGIs, on the other hand, only make contact with the MoGs in segments 1–3 (asterisks indicate the absence of a contact), the result being that half of the tail contracts and the animal moves forwards. (Data from Wine, J. & Kranse, F.B. (1982) The cellular organisation of crayfish escape behaviour. In *The Biology of Crustacea*, ed. by Bliss, D.E. *et al.*, pp. 241–92, Academic Press, London. Reproduced with permission of Elsevier.)

and role of the LGI component of the tail-flip control system has been particularly thoroughly researched and is therefore the component that we will now concentrate upon.

The LGI circuit

Sensory input to the LGI system is gathered from the environment via the displacement of approximately 1000 cuticular hairs (termed sensory afferents) covering the abdomen of the animal. These hairs operate in manner broadly similar to the spider trichobothria that were considered as a case study earlier in this chapter. Their mechanism is more similar to that of the insects than to that of the spider however, in that a bipolar cell that is sensitive to direction of movement serves each hair. The bipolar cells communicate with the LGI either directly or via a bank of 25–50 sensory interneurons. Each abdominal segment contains a single LGI, and these connect to one another to provide a chain for the transmission of signals from the tail all of the way up the animal's dorsal nerve cord.

Output from the LGI ultimately causes the rapid contraction of the abdominal fast flexor muscles, flexing the abdomen as shown in Fig. 2.11. Two routes achieve this. The first is via direct excitation of motor giants (MoGs). These neurons form a direct contact between the LGI and all of the fast flexor muscle fibers in each abdominal segment. The second route interposes an interneuron called a segmental giant (SG) between the LGI and a set of fast flexor neurons (FFs), each of which connects to a few of the fast flexor muscle fibers within a segment.

Figure 2.12 shows the arrangement of the various components of the LGI-mediated tail-flip system both in the tail of a crayfish and as a simplified circuit diagram to explain its connectivities. It should be obvious from the circuit diagram that the LGI has a special and central role in the system. Physically it represents a convergence of inputs from a large number of sensory cells and it has an effect upon a wide range of motor cells. This crucial position within the system identified the LGI as a key candidate to be the command neuron or "on/off switch" of the tail-flip response. But being in the right place both physically and in the signal sequence does not actually prove that the LGI is the command neuron. In order to do this we need to consider three further lines of evidence. Firstly, the LGIs can be described as being both sufficient and necessary for the tail-flipping to take place. Simple direct electrical stimulation of a giant interneuron will result in

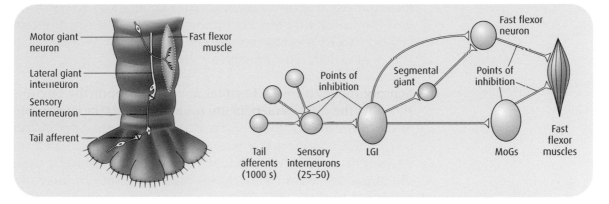

Fig. 2.12 The physical and schematic LGI circuit. Note the point of input/output convergence represented by the LGI itself. Points of LGI command-derived inhibition are indicated by arrows. (After Carew, T.J. (2000) *Behavioral Neurobiology*. Sinauer Associates, Sunderland, MA.)

the performance of a tail-flip. If the LGI is stimulated in this way the result will be an LGI-type flip. This demonstrates the sufficiency of the LGI in this context.

Demonstrating necessity is slightly more difficult. It is one thing to say that the LGI can cause a tail-flip in an experimental situation, but under natural conditions a tail-flip could be possible in the absence of a functioning LGI because some bypass or parallel circuitry exists. It is possible to switch off (temporarily) the intact LGI of an animal by hyperpolarizing it so that no level of excitatory input from the sensory afferents can persuade it to spike. Under these conditions no tail-flips are observed and so it seems unlikely that parallel or bypass systems exist.

The second area of evidence is based upon the idea that the LGI is the decision-maker in the system. We know that the LGI is necessary for a tail-flip, but can we assume that it controls the decision to flip? If it does not we should be able to identify a decision-maker upstream of the LGI in the neural circuit, i.e. in the bipolar sensory cells themselves, or in the sensory neurons that they signal to. Careful monitoring of the output of these components of the system in response to gradually increased levels of external stimulation demonstrates that the LGI has a threshold of input that must be reached before it can generate a spike. This observation, coupled with the evidence already presented regarding the sufficiency of LGIs, explains the fact that the tail-flip is an all or nothing behavior. The final piece of evidence to secure the command neuron status of the LGI comes from the observation that in addition to controlling the onset of tail-flipping, it also controls the inhibition of a range of other behaviors that would be incompatible with a flip.

LGI command-derived inhibition

When the LGI achieves its action potential a tail-flip is initiated. At the same time vitally important inhibition in the same circuit is elicited to ensure that a second flip cannot be performed (in response to continuing input from the sensory afferents) until abdominal re-extension has taken place. The various points in the circuit at which LGI-mediated inhibition occurs are indicated in Fig. 2.12. Postsynaptic inhibition of the sensory neurons and presynaptic inhibition of the sensory interneurons prevent the transmission of information to the LGIs. The LGIs themselves are inhibited to prevent the production of a chain of spikes and flips, and the MoGs are inhibited (as soon as they have had time to innervate their target muscles only once) to prevent the continued flexion of the abdomen.

Key reference
Edwards, D.H., Heitler, W.J. & Krasne, F.B. (1999) Fifty years of a command neuron: the neurobiology of escape behavior in the crayfish. *Trends in Neuroscience*, **22**, 153–61.

Post-flip re-extension

When the tail-flip is complete it is necessary for the abdomen to re-extend. The LGI is not directly involved in this stage of the behavior, but it should be remembered that it can only occur once the command-derived inhibition of the LGI system subsides. Flexing of the abdomen during the flip stretches its muscle fibers. This causes a sensory neuron (a stretch receptor, or muscle receptor organ, MRO) to fire. The MRO communicates the fact that the abdomen is stretched to the central nervous system. It has an excitatory connection with fast extensor motor neurons and an inhibitory connection (via an interneuron) with the abdominal fast flexor muscles. The activity of both of these connections is combined with direct input to the extensor motor neurons from hair receptors on the abdomen to elicit re-extension.

So it would appear that the crayfish has an elegantly simple and highly stereotypical behavioral response to the approach of a predator. This is the case, but it would not be true to say that the tail-flip is always produced when the abdomen is tapped. We will return to this observation in Chapter 4 when we consider the adaptive significance of variability in the performance of this behavior.

Summary

This chapter has introduced some of the basic components of the machinery by which the nervous system exerts a level of control over the expression of behavior. Through the examination of a relatively small number of thoroughly researched examples a number of specific points have been made. Firstly, that apparently simple reflex behaviors provide a useful model through which to investigate aspects of the control of behavior. Stimulus filters and feature detectors allow an animal to gather crucial information from its environment while ignoring the "noise" in its system, while dedicated systems and command neurons maximize the efficiency of nervous communication. Nervous coordination is often achieved through a carefully controlled balance of stimulation and inhibition of key nervous connections. Finally, there is an element of flexibility in apparently rigid behaviors to allow an animal to compensate for variability in its environment.

Questions for discussion

In this chapter we have only considered mechanical and visual sensory input. What other forms of sensory input might animals use, and in what circumstances would each of these senses seem to be most appropriate?

Do you think that some classes of behavior are more likely to come under the heading of "reflex" than others and why did you reach this conclusion?

This chapter closed with the observation that a crayfish will not tail-flip every time that its tail is tapped. Why do you think that this should be the case? (You should consider this question before we go on to consider this in Chapter 4.)

Further reading

Neurobiology is a vast and fascinating subject in its own right and in this chapter we have really only begun to scratch the surface. *Nerve Cells and Animal Behavior*, by Peter Simmons and David Young (1999, Cambridge University Press, Cambridge) provides a useful overview of the breadth of the topic and is an effective entry point to the wider literature. *Behavioral Neurobiology* by Thomas Carew (2000, Sinauer Associates, Sunderland, MA) is an excellent introduction to both the breadth and depth of the subject; it concentrates on a few case studies but thoroughly investigates every aspect of them, and will excite any student of behavior.

3 The Motivation and Organization of Behavior

Time is nature's way of keeping everything from happening at once.

Woody Allen

In Chapter 2 we discussed the way in which the nervous system exerts a level of control over the behavior of an organism. We saw that one class of behaviors, the reflexes, could be thought of, quite literally, as a knee-jerk response to a stimulus. However, we also noted that the control of some behaviors was far more complex than this and that their performance may depend upon a range of factors both internal and external to the organism concerned. In this chapter we will explore this further by considering the factors that motivate an individual to perform a particular behavior at a particular time.

Key points

• Behaviors can be thought of as having underlying motivations. These may be internal or external to the individual (or both). Levels of motivation relating to homeostasis are regulated in accordance to need by a range of physiological mechanisms. Variations in motivation tend to increase fitness (see "Focus on fitness and coefficients of relatedness" in Chapter 4).

• Many behaviors are expressed according to a distinct temporal pattern. They occur with a regular periodicity and are controlled by biological clocks.

• There is a genetic component to the control of biological rhythms, but their cycling is also controlled to some extent by internal and external environmental cues.

Motivation

Homeostasis and the motivation to drink

Imagine the following situation. You are standing on a sandy beach on a very hot day. Being an athletic type you have just played a particularly gruelling game of beach volleyball. What do you think you might feel? Well, depending upon the outcome of the match you may be experiencing the thrill of victory or the gloom of defeat. But at a more basic level I should think that you feel very hot and very thirsty.

Our bodies have a built-in system of regulators that maintain our internal environment at the peak of operating efficiency. In response to your activity, the heat of your surroundings, and the heat that your muscles have generated, your body's regulator system will "tell" you to cool down. You probably felt that you wanted to move into the shade or take a cooling swim, i.e. to behave in a way that would facilitate cooling. But the game came first and so you didn't. In which case your regulators stepped in and took charge of the situation themselves. You will have started to sweat, and blood will have been pumped into the capillaries close to the surface of your skin. This will have enabled your body to loose some heat, but of course it has also resulted in water loss, and that explains your thirst.

In very anthropomorphic language (of course your body doesn't actually "tell" you things and it doesn't "take charge" in the sense that you or I might take charge of a situation) I have just described an aspect of the phenomenon that is homeostasis. There is a tendency to regard the homeostatic process as a set of mechanisms whereby sensors record a change in the body's internal environment and systems are activated in order that the status quo is maintained. But think again about our hypothetical day at the beach. If you are an experienced beach volleyball player you might very well have anticipated the dehydration that the game would cause, and drunk extra water before the match. Your prior experience might have had an effect upon your drinking behavior. The important point to be made here is that whether your drinking was anticipatory or responsive, your condition caused a change in your behavior. You made the transition from not drinking to drinking. Another way to put this would be to say that you were **motivated** to drink, or that your **motivation** to drink changed. Motivation is a key concept in animal behavior

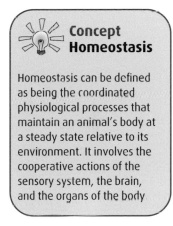

Concept Homeostasis

Homeostasis can be defined as being the coordinated physiological processes that maintain an animal's body at a steady state relative to its environment. It involves the cooperative actions of the sensory system, the brain, and the organs of the body

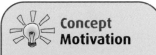

Concept Motivation

In human terms we think of motivation as a desire to do something. We would probably make a distinction between calculated rational motivations with a conscious goal and more impulsive motivations driven by an internal urge such as hunger or fear. In fact we should probably consider these to be the extremes of a continuum of motivational types.

and one that is clearly related to variation in the behavior of an individual. We can therefore think of motivation as being part of a decision-making process.

The loss of water from the body of a mammal is recorded by specialist cells termed osmoreceptors that are located in the hypothalamus of the brain. Changes in the internal water balance of the body (as a result of drinking, or dehydration due to excretion, sweating, respiration, etc.) alter the osmotic environment of the cells. A water deficit results in the net movement of water out of the cells and into the intracellular matrix, or into the blood stream. This water loss causes cell shrinkage, as turgor is lost. It seems likely that the osmoreceptors are sensitive to their own volume changes and use their shrinking as a cue to both initiate hormonally controlled water conservation measures (e.g. they stimulate the release by the pituitary gland of antidiuretic hormone which decreases urine production by the kidneys), and to motivate the onset of drinking behavior. This is an example of a **negative feedback** system, i.e. one in which the level of resource seeking behavior (drinking) increases because the levels of the resource reserve (body fluid) have decreased.

However, if we return briefly to our volleyball player we have seen that the factors controlling motivation to drink are unlikely to be as "simple" as a physiological response to fluid balance. After all an experienced athlete might anticipate the dehydration that is likely to result from their exertions and drink plenty of water before there was a physiological need to respond to. In a similar way we now know that a range of animal species can use cues other than actual dehydration to trigger drinking behavior. Both rats and pigeons will increase their drinking levels in response to a rise in air temperature even though their internal fluid balance has not in fact changed. There can therefore be a behavioral component involved in homeostasis, both in response to current need and in anticipation of a future requirement. This could imply that animals learn to associate warmer conditions with a dehydration risk on the basis of past experience, that they remember this, and that they act accordingly. We will return to this idea in Chapter 4 when we consider learning as a concept in more detail.

The motivation to eat

If the motivation to drink can be a result of either physiological need or learned anticipation of need then we would also expect a

Link
Animals can learn how to satisfy their homeostatic needs.
Chapter 4

range of factors to motivate feeding behavior. If I were to ask you "how do you know when to eat" your most likely reply would be, "when I am hungry". So hunger must motivate feeding behavior. In Chapter 6 I will ask the same question again (and add to it how do you know what to eat and how much of it to eat) and we will see that a whole range of factors other than hunger per se are involved in foraging decisions. But I do want to consider feeding and the motivation to feed in a little more detail in the current chapter.

So what is **hunger**? Dictionaries define it as "a strong desire to eat" and at a physiological level it has recently been established that the actions of the hormones leptin and ghrelin are key regulators of the desire to eat, or appetite, in humans and a range of other mammals. So we could think of hunger as being the response of the organism to a hormonal signal. Leptin is released by adipocytes, specialized fat storage cells. Increased fat storage results in high levels of circulating leptin and so leptin levels can be used as an indicator of the adequacy of the body's stored energy reserves. Three areas of the hypothalamus, the arcuate nucleus, the lateral hypothalamic area, and the paraventricular nucleus, monitor the levels of leptin in the body. If leptin levels are high appetite is suppressed. If leptin levels are low the stomach secretes the hormone ghrelin. This acts upon the same three areas of the brain but its effect is to stimulate appetite, i.e. to signal hunger. So there we have it, in physiological terms the motivation to eat could be said to be a reaction to high levels of the hormone ghrelin.

Hyperphagic behavior

The relationship between body energy reserves and appetite as expressed through feeding behavior can clearly be seen in the results of experiments carried out by Neil Metcalfe and Colin Bull of the University of Glasgow, Scotland. They induced varying degrees of hunger in juvenile atlantic salmon *Salmo salar* by starving them for varying periods of time. One group of fish were deprived of food for 40 days (very hungry) and a second group were starved for 20 days (hungry). As would be expected, the members of the former group used up more of their bodies' fat reserves than did the latter and could thus be considered to be the hungrier of the two. Or to put it another way we could say that the fish starved for the longer period had the greater energy deficit. I should point out at this stage that these apparently long periods of food deprivation were not unduly harsh. These experiments were carried out during the winter months when wild juvenile salmon frequently face long periods of limited food supply. In fact it was exactly because of this situation that the experiments were carried out. The researchers wanted to know how the fish responded to this situation behaviorally. Their hypothesis was that the hungry fish would be motivated to replenish their bodily fat reserve by exhibiting hyperphagia (literally overeating) and that hungrier fish would have the biggest appetite and the strongest hyperphagic response. One would assume that the fish could increase the intensity of their hyperphagia (eat more, more quickly) or the duration of their hyperphagic behavior (eat more for longer), or that they could do both.

After the experimental starvation period the fish were provided with food and their feeding behavior was monitored. The two groups were compared to one another, and both were compared to a control group composed of fish that had not been starved (i.e. would demonstrate a "normal" hunger level). Given that the fish in both of the experimental groups were hungry they would both be expected to eat more than the members of the control group, and that is exactly what the results showed (Fig. 3.1). The very hungry fish and hungry fish exhibited similarly high levels of hyperphagia at the onset of refeeding but the very hungry fish maintained higher levels of feeding behavior for longer than the hungry fish. This suggests that starvation resulted in a strong motivation to feed in both groups, but that as a result of their higher energy deficit the very hungry fish were motivated to feed at higher levels for a longer period. If you recall, the original hypotheses were that the very hungry fish would demonstrate either a higher intensity of hyperphagia or a longer duration of hyperphagic behavior than the hungry fish. The data clearly support the latter, but why is it that the fish do not seem to do the latter? If you want to answer this question I suggest that you skip forward to Chapter 6.

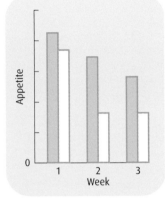

Fig. 3.1 Changes in appetite of very hungry ▣ and hungry ☐ fish relative to an unstarved control population (0 level) during the first 3 weeks of re-feeding. (From Bull, C.D. & Metcalfe, N.B. (1997) Regulation of hyperphagia in response to varying energy deficits in overwintering juvenile Atlantic salmon. *Journal of Fish Biology*, **50**, 498–510. Reproduced with permission of Elsevier.)

Motivation and nonhomeostatic systems

Thus far we have considered examples of motivation that have been related to internal "needs" and stimuli and through them to the regulation of an aspect of the physiology of the individual. It would be remiss of me to allow you to progress to the next section of this chapter thinking that motivation is only related to homeostasis. It is not. For example experimental observations of female rhesus monkeys *Macaca mulatta* have demonstrated that sexual motivation (measured as the lengths that a female would go to to achieve a mating) is highest at a particular period of her estrus cycle. The closer she is to ovulation, the more highly sexually motivated she is. This is largely due to the relative levels of the sex hormones testosterone, estrogen, and progesterone circulating in her blood. So the monkey's motivation to mate and to eat are similar in that there is a hormonal component to both and both are a result of the animal's internal state. But they are dissimilar in that one motivation (feeding) is related directly to a physiological balance and therefore to the homeostatic system while the other is not.

It would also be wrong of me to leave you with the notion that motivation is predominantly concerned with internal stimuli such as hormones. Clearly these are very important, but as I have already mentioned a rise in air temperature can motivate drinking behavior in a range of species. The sexual motivation of male rhesus monkeys also increases in phase with that of the female, to a peak that coincides with the female peak. Male testosterone levels are known to increase male sexual motivation, but the fact that the motivation of the male reaches its peak at a time coincident with that of the female and therefore with female ovulation suggests that male motivation is also dependent upon external stimuli derived from the female.

We will consider the effect of a range of stimuli, both internal and external, upon a wide range of motivations throughout the remainder of this book.

Biological rhythms: clocks and decision making

Rhythms and feeding

We have already discussed the way in which an increased motivation to feed causes an increase in feeding behavior when food is presented to previously starved salmonid fish. You will remember that in their natural environment these fish have to contend with a regular period of winter starvation and that they follow it with a period of compensatory hyperphagia. We could describe this by saying that the animals experience a repeating cycle of food

> **Concept**
> **Variations in motivation increase fitness**
>
> At times when an animal's fitness could be increased motivation tends to increase. For example, reproductive fitness is increased by successful matings and sexual motivation tends to increase in a way that enhances the chance of reproductive success.
>
> Changes in motivation related to homeostasis maintain the efficient functioning of the body, increasing the probability of survival. They therefore have important consequences in fitness terms.

shortage and food abundance. Or to put it another way we could say that the animals describe an annual cycle of winter fasting and springtime gluttony. Careful observations of the minute by minute feeding behavior of feeding salmonids reveals the existence of another cycle of feeding with a far shorter periodicity.

Francisco Sánchez-Vázquez and Mitsuo Tabata have carried out experiments to record the daily rhythms of feeding behavior exhibited by the rainbow trout *Onocorhynchus mykiss*. They provided individually housed fish with an apparatus that allowed the animals to regulate their own food delivery and therefore their own food intake. The fish were first trained to operate this "demand-feeder" that discharged a pellet of food every time the fish activated a tactile sensor. In effect the fish was trained to press a button to gain a food reward in a modification of the Skinner-box apparatus that we will encounter again in Chapter 4. Of course if the fish were starved we would expect them to exhibit hyperphagia and to feed constantly, but they were not starved and by monitoring their feeding pattern the researchers were able to gain insights into their short-term feed/not-feed decisions. Figure 3.2 is an actogram, a pictorial record of the feeding activity of one fish.

Initially the fish were maintained under a 12-hour light and 12-hour dark cycle (12 : 12 h LD), and for the first 35 days of the experiment it is clear that feeding activity was largely restricted to the 12 light hours of the day. We would describe this as a circadian rhythm.

Fig. 3.2 An actogram of the feeding activity of a group of trout kept under 12-hour light and 12-hour dark conditions from day 1 to day 35 and exposed to 45-minute pulses of light and dark from day 35 to day 50 of the experiment. (From Sánchez-Vázquez, F.J. & Tabata, M. (1998) Circadian rhythms of demand-feeding and locomotor activity in rainbow trout. *Journal of Fish Biology*, **52**, 255–67. Reproduced with permission of Elsevier.)

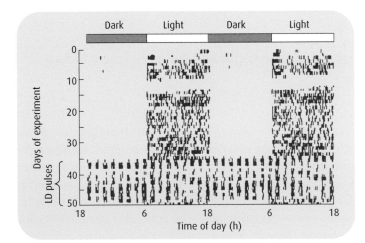

A possible explanation for the rhythm might relate to hunger levels and the motivation to feed. Perhaps the fish eat for 12 hours, become satiated, and are hungry again 12 hours later, their hunger initiating another 12-hour-long feeding bout. But look what happened during days 35–50. The neat rhythm has disappeared and in its place the fish demonstrate activity and inactivity throughout the 24-hour period.

You should notice that this change in behavior coincides with a change in the environment of the animals. The 12 : 12 h LD lighting regime has been replaced with short alternating pulses of light and dark of just 45 minutes' duration (45 : 45 min LD). This makes our hypothesis that the fish eat their fill and get hungry again 12 hours later seem highly improbable. It seems more likely that the bouts of feeding are in some way linked to the periods of light and dark. Under both of the experimental illumination regimes that we have thus far considered the fish restrict their feeding to the period of light.

So what happens when the lights are left on constantly? The fish retain their circadian rhythm, in that they feed during an approximately 12-hour period and they fast during a period of similar length. But the onset of these periods is a little later with each day of the experiment, it seems to drift as the days of continual light progress. To describe this phenomenon we would say that the rhythm of behavior was free-running. (This is still a circadian rhythm because it retains a c. 24-hour periodicity.) Crucially a return to a 12 : 12 h LD environment restores the behavioral rhythm so that it is **in phase** with the rhythm of illumination. The explanation for these observations is as follows. The fish have an internal (or endogenous) rhythm of feeding activity with a periodicity of approximately 24 hours (in fact 25.8 hours), i.e. they demonstrate a circadian rhythm. This rhythm is "kept on track" by an external cue termed a *zeitgeber*. This German word translates as "time-giver" and we should think of it in exactly those terms. In effect it "tells" the animal when it is time to initiate the behavior and so sets the rhythm. In this case the zeitgeber is the turning on and off of the lights. This is why the very regular 12 : 12 h LD cycle resulted in a very regular 12-hourly cycle of feeding activity and inactivity and why the rapid 45 : 45 min LD shifts of light and dark resulted in similarly short pulses of feeding and nonfeeding. Kept under constant light the animals had no external cue by which to set, or **entrain**, their internal clock. In the absence of the zeitgeber the endogenous rhythm ran to its

**Concept
Circadian
rhythms**

The term circadian rhythm describes a a pattern of behavior that repeats itself on an approximately 24-hour cycle. The term is derived from the Latin words *circa* and *dies* which translate as "about a day."

Key reference
Sánchez-Vázquez, F.J. & Tabata, M. (1998) Circadian rhythms of demand-feeding and locomotor activity in rainbow trout. *Journal of Fish Biology*, **52**, 225–67.

internally derived periodicity of 25.8 hours rather than the externally cued periodicity of 24 hours, and so we record the drift.

Circannual rhythms

Many species of animal demonstrate periods of behavior with a one-yearly repeating cycle. For example, the passerine birds that migrate into the northern temperate regions of the world to breed exhibit an annual cycle of testicular growth and regression and demonstrate regular molting cycles that have been shown experimentally to be under endogenous control. During the colder winter months these birds return to their southern wintering grounds, but migrations of this kind are not an option for all temperate animal species. Those that cannot remain active in the face of the adverse conditions that winter brings must adopt a different strategy. Many undergo an annual hibernation that is often associated with prehibernation hyperphagia and shelter-seeking behaviors such as burrow or roost site preparation. Experimental work has confirmed that this phenomenon too is under endogenous control (Plate 3.1).

Longer and shorter rhythms

The performance of a great many behaviors demonstrates the existence of a clearly defined rhythm that cannot be described as being either circadian or circannual. Those that have a cycle shorter than 24 hours are termed **infradian** rhythms and those

Link
The onset of migration is controlled by a biological clock.
Chapter 4

Plate 3.1 The autumnal burrowing behavior of this Mediterranean tortoise is under endogenous control. © G. Scott.

having a periodicity that is longer than a day are called **ultradian** rhythms.

For example the environment of intertidal animals changes dramatically with the twice-daily ebb and flow of the tide. Not surprisingly many of the behavioral rhythms of these animals are in phase with the tidal cycle, having a periodicity of a little over 12 hours. Rhythms of activity with a periodicity ranging from a few minutes to about 40 minutes have been recorded from a number of diurnal bird species with respect to their feeding, drinking and preening behavior, although of course these rhythms will overlie the circadian rhythm of daytime activity and night-time inactivity.

Case study The circatidal swimming of the shanny

Intertidal fish face an obvious problem when the tide falls. Their intertidal habitat becomes dry land and is therefore no longer available to them. Some species retreat with the tide and return when it rises. Others such as the shanny (*Lipophrys pholis*) take refuge in a rock pool to avoid becoming stranded on dry land. Not surprisingly shannies are most active when the tide is in, and if they are removed to constant laboratory conditions their activity cycle can be preserved for a number of days, although as the days progress the rhythm does dampen and lose precision. The peaks of their swimming activity tend to coincide with the expected high tide and the cycle has a mean periodicity of 12.4 hours. These fish clearly demonstrate an endogenous circatidal rhythm (Fig. 3.3).

After prolonged exposure to constant conditions the fish become arrythmic, but if the fish are replaced on the shore and exposed to just a few tidal cycles their rhythm will be restored. Moreover the restored rhythm will be in phase with the tidal cycle of the shore that they are returned to, even if this is not their shore of origin (Plate 3.2).

Replacing the fish on the shore exposes them to a whole range of potential stimuli that could have the clock-setting role. So in order to pinpoint the zeitgeber involved arrythmic fish have been exposed to a variety of individual environmental components. Cycles of feeding, temperature change, current, and water chemistry,

all of which are commonly associated with tidal change, do not result in the restoration of the rhythm. But two factors do seem to be important. The rhythm returned when the fish were caged on the seabed, but not when

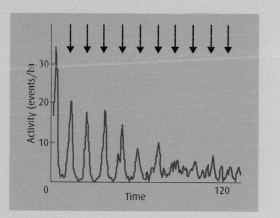

Fig. 3.3 The amount of swimming activity of shanny that have been moved to a tide-free environment. Arrows indicate positions of high tides in the natural environment of the fish. (From Morgan, E. & Cordiner, S. (1994) Entrainment of circa-tidal rhythm in the rock-pool blenny *Lipophrys pholis* by simulated wave action. *Animal Behaviour*, **47**, 663–9. Reproduced with permission of Elsevier.)

Plate 3.2 Like the shanny, this sea scorpion has a rhythm of activity linked to tide state. © C. Waller.

the cages were allowed to float just under the water surface, i.e. the cycle of changing water depth and therefore hydrostatic pressure seems to be important. Experimental manipulations confirmed this to be the case, and in fact hydrostatic pressure has now been identified as being perhaps the most effective phase-setting and entraining agent for the rhythms of intertidal fish. The second factor found to be important in the entrainment of the rhythm is wave action. The simulation of wave action for 2 hours at 12-hourly intervals will result in the restoration of the activity rhythm (Fig. 3.4).

It seems likely that in the field hydrostatic pressure and wave action act together in an additive fashion because exposure to just one or two real tides is sufficient to entrain the rhythm. By contrast, 11 or more exposures to 12-hourly pulses of either hydrostatic pressure or wave action alone are required in the laboratory. Of course it may be that other as yet unidentified tidal variables also have a part to play.

Morgan, E. & Cordiner, S. (1994) Entrainment of a circa-tidal rhythm in the rock-pool blenny *Lipophrys pholis* by simulated wave action. *Animal Behavior*, **47**, 663–9.

Northcott, S.J. (1991) A comparison of circatidal rhythmicity and entrainment by hydrostatic pressure cycles in the rock goby, *Gobius paganellus* (L.) and the shanny, *Lipophrys pholis* (L.). *Journal of Fish Biology*, **39**, 25–53.

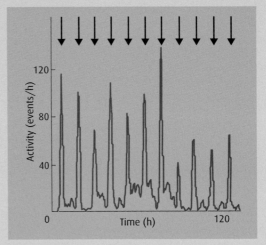

Fig. 3.4 The amount of swimming activity of shanny subjected to artificial wave conditions to simulate high tides at 12-h intervals (arrows). (From Morgan & Cordiner 1994. Reproduced with permission of Elsevier.)

Rhythms having a periodicity of more than a day are also common. Human females, for example, have a menstrual cycle of 28 days and human males have a 21-day cycle of testosterone release. Both of these cycles are likely to have an effect upon the behavior of the individual. Many organisms exhibit rhythms with a periodicity of around 29.5 days in phase with the lunar cycle. Perhaps the most famous of these is the breeding cycle of the marine palolo worm. These animals are reproductively active only on those nights of the last quarter of the moon during the months of October and November.

Perhaps the most impressive biological rhythms are those concerning the emergence behavior of the various species of periodic cicada. The nymphs of these insects live underground where they mature over a number of years, feeding on the contents of root xylem. The exact length of time they take to reach maturity depends upon the species concerned, the 17-year cicada *Magicicada septendecium* takes 17 years. On maturation the whole population emerges and as adults they live for just 2 weeks, during which time they reproduce. Exactly how the cicadas keep track of time has always intrigued researchers and it has always been assumed that the insects must rely on an internal clock. Recently however one group of scientists working with the 17-year cicada in California have suggested that the nymphs use an external cue and that they can count. For their experiments they took 15-year-old nymphs and moved them to an experimental enclosure. These nymphs should have taken a further 2 years to emerge as adults, but in fact they took just 1 year. The researchers had made this happen by lengthening the period of daylight to which the peach trees on whose roots the insects fed were exposed. By doing this the trees were "tricked" into flowering twice during the year rather than the usual once. Flowering in trees coincides with a peak in amino acid concentrations in the sap that the insects feed on. So it seems that the cicadas "tick-of" the years by counting the peaks.

Controlling the rhythm

Experiments that are designed to demonstrate the existence of endogenous rhythms and to identify the zeitgeber that entrains them can only tell us so much about the processes underlying clock function. Some of the key questions that have exercised scientists working in this area of animal behavior concern the physical location of the clock within the body of the animal, the pathway by which entrainment occurs, and the internal mechanism by which rhythms are generated and maintained.

There is a genetic component to behavioral rhythms. The activity of a single gene is essential for the normal expression of the circadian locomotory rhythm of mice. A mutation in the gene, that has appropriately been named *clock*, will result in the loss of the normal rhythm, which has a periodicity of a little less than 24 hours when it is allowed to free-run in constant (dark) conditions. The mutant mice have a rhythm that free-runs on a 28-hour cycle initially, but breaks down completely after a period of a few days.

The fact that the mice I have just described were maintained in constant darkness to allow their rhythm to free-run is significant. The zeitgeber in this case is the light/dark cycle that the animal is exposed to. Mice receive information about light levels via their optic system, and so if the biological clock does have a physical location the visual system would seem to be an obvious place to look for it. Blinding mice by severing the link between the retina and the optic nerve does result in free-running rhythms and so the eyes must be the starting point of the entrainment pathway. Severing the optic nerves at progressive distances from the retina still results in free-running rhythms until the optic chiasma (the point that the nerve from the right and left eyes cross one another into the left and right hemispheres of the brain respectively) is reached. Beyond this cutting the nerves does result in visual blindness, but the animals must still be light/dark sensitive because their circadian rhythms persist. In fact projections from the retinal ganglia exist that make direct connections with the hypothalamus in a "forward" position relative to the optic chiasma. This link between the retina and the brain is termed the retinohypothalamic tract or RHT. The hypothalamus is known to be involved in the regulation of a wide variety of behaviors and so it was perhaps not surprising that the entrainment pathway should lead there. The RHT ends at a specific pair of hypothalamic nuclei referred to as the SCN (suprachiasmatic nuclei). In those mammals that have undergone experimental destruction of the SCN the expression of circadian rhythms of behavior is permanently lost. It would seem therefore that the SCN is the physical location of the mammalian biological clock, and that of other vertebrates. (In fact the genetic components of the clock are the same across the animal kingdom.)

The light sensitivity of the SCN appears to be linked to variations in the levels of activity of a few key genes such as *per* (period) and *tim*. The protein products of these genes interact with one another and their relative concentrations in the cells of the SCN vary in a way that mirrors the behavioral rhythms of the organism. It is assumed that these protein concentrations translate into signals from the SCN to the pineal gland (whether these are neuronal or hormonal remains to be seen). This stimulates the gland to secrete melatonin, a hormone that is known to play an important part in the control of both physiological and behavioral rhythms (Fig. 3.5).

Link
Per is a pleiotropic gene that has a number of effects.
Chapter 4

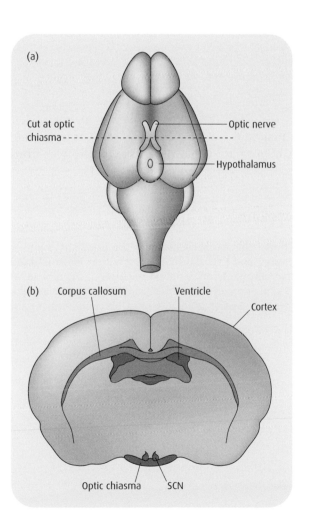

Fig. 3.5 (a) The ventral surface of a rat brain, showing the position of the optic nerves, optic chiasma, and the outline of the hypothalamus. (b) A section of the brain made at the point labeled "cut" in (a), showing where the paired SCN lie just above the optic chiasma. (From Toates, F., ed. (1998) *Control of Behaviour.* Springer-Verlag, Berlin. Reproduced with permission of Springer-Verlag.)

Summary

This chapter has introduced the concept of motivation. It has examined the ways in which the performance of a particular behavior might depend upon one or more cues or stimuli that may be either internal or external to the individual. Some of these behaviors have no particular periodicity and occur as and when the need arises; others though follow a distinct rhythm, under the control of an internal clock. The basis of these clocks is not yet fully understood, but some animals appear able to mark time, and some clocks have a clear genetic basis.

Questions for discussion

Bull and Metcalfe found that juvenile Atlantic salmon increase the period of their hyperphagia in response to periods of increasing starvation. However, they do not appear to increase the intensity of their feeding in response to increased starvation. Why not? Skip ahead to Chapter 7 to find some possible answers.

Are your own patterns of work and play controlled by your internal clock? To what extent are you "slaves to the rhythm"?

Further reading

Animal Motivation by Patrick Colgan is an excellent book, dealing with both the internal and external mechanisms that control animal behavior. If you want an absolutely up to date discussion of the biological rhythms of animals (including humans) read *The Living Clock: The Orchestrator of Biological Rhythms* by John D. Palmer (2002, Oxford University Press, Oxford).

4 The Development of Behavior

The Stork in the sky knows the time to migrate.

The Book of the Prophet Jeremiah, chapter 8, verse 7

The previous chapter introduced the idea that there could be a genetic component to the control of behavior. However, it should also be clear to you that your own behavior and that of the animals around you will vary with, among other things, experience. In this chapter I want to explore both of these areas further and at the same time highlight the integration that must exist between both of them.

Key points

• Genetic variability is linked to variability in expressed behavior, but this does not mean that behavior is entirely under the control of genes.

• The particular patterns of behavior expressed by an individual will depend upon an interaction of the animal's genotype and both the internal environment in which those genes are expressed, and the external environment in which the animal lives.

• Some behaviors are innate, but many are modified as a result of experience, i.e. behaviors can be learned.

• Animals make use of a wide variety of cues to enable them to successfully navigate their environment and to undertake feats of migration.

Genes and behavior

Genetics, a brief refresher

It is beyond the scope of a book such as this to discuss genetics in any detail, and so I will assume that the majority of readers will have a basic grasp of the terminology and processes involved in the transmission of genetic material from generation to generation and of patterns of heredity. If you do not it would not be a waste of your time were you to read an introductory textbook on the subject. In the context of this chapter I think that it is enough to remind ourselves that chromosomes consist of a double helix of DNA and that specific sections of this DNA act as templates for the production of proteins. These sections of DNA are termed **genes** and their specific sites on the chromosome are termed **loci** (not all proteins are the product of a single gene but we really do not need to worry about that in this context). Each individual gene is always at the same **locus** and it always codes for the same protein. This means that it also always affects the same character in the organism, be that an aspect of behavior, physiology, or growth form.

There is however an element of variation possible within individual genes that translates into variation in the phenotypes affected by them. For example, in humans the gene for eye color exists as a number of variants each of which results in a different eye color. The different forms of the gene are termed **alleles**. So possession of one allele will result in blue eyes whilst possession of another produces eyes that are brown.

Except in those specific cases that involve asexual reproduction, an organism will have inherited one set of chromosomes from its mother and another set from its father. It will therefore have two alleles at each locus. If both of these alleles are identical the individual is described as being **homozygous** at that locus. If however it has inherited two different alleles at a locus it is referred to as being **heterozygous**.

Because a homozygous individual has two copies of the same allele it can only possibly exhibit one phenotype. But what happens in the case of a heterozygote? In fact there are two possible outcomes in this scenario. In some cases the expressed phenotype is an intermediate between those normally expressed by individuals that were homozygous for each of the alleles involved. However, in other cases the phenotype determined by one allele would be expressed and that of the other would not. So for example if you

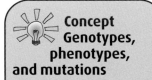

Concept
Genotypes, phenotypes, and mutations

The genetic complement of an individual is referred to as its genotype. The word phenotype describes the biochemical, physiological, morphological and behavioral characters of the individual. Phenotype is in part determined by genotype, but it is also strongly influenced by environmental conditions.

Mutations in the genotype (changes in the make-up of genes) can result in changes in phenotype.

have blue eyes it is certain that you are homozygous for the blue allele. If you have brown eyes, however, you could be homozygous for the brown allele, or you could be a heterozygote having one copy of each allele. In such a case it would be impossible to determine you genetic make-up just by gazing into your eyes.

The reason for this variation is that brown alleles code for a pigment that makes the iris of the eye brown. Blue alleles, however, lack the ability to produce the pigment and it is actually a lack of brown that makes blue eyes appear blue. It only takes one copy of the brown allele for the expression of the brown eye phenotype to be possible.

When one allele in a heterozygote results in the expression of its usual phenotype, but the other does not, we would refer to the expressed allele as being **dominant** and to the allele that is not expressed as being **recessive**.

Single gene effects, foraging flies, and hygienic bees

Larval fruit flies spend most of their time foraging. They crawl through their food shoveling yeast into their mouths with feeding hooks. In the laboratory of Marla Sokolowski there are larval *Drosophila melanogaster* that exhibit two very different foraging behavior phenotypes. When allowed to feed on a yeast covered petri dish the larvae of one population will move considerable distances (around 8 centimeters which is a long way for a tiny maggot) in a 5-minute test period, whereas the larvae of another population will hardly move at all. The latter are known as "sitters" and the former as "rovers". Flies exhibiting both of these phenotypes or strategies are found in the wild where they occur with frequencies of approximately 70% rover and 30% sitter.

Those of you who are familiar with basic patterns of genetic inheritance should find these frequencies, a 3 : 1 ratio, highly suggestive of a simple Mendelian pattern reflecting a system that involves a single gene with one dominant allele and one recessive allele. Through a program of carefully controlled crosses Sokolowski and her coworkers have confirmed that this is the case.

Their results, a simplified version of which are presented in Fig. 4.1, identify a single gene which they named *foraging* that has two forms, the dominant *for*^{R} (rover) and the recessive *for*^{s} (sitter). Flies homozygous for *for*^{R} therefore exhibit the rover phenotype as do heterozygote flies. Those individuals homozygous for *for*^{s} exhibit the sitter phenotype. The researchers have even been able

Concept Genes for behavior

There is a tendency to think that there are genes **for** particular behaviors. Whilst it is true that genetic differences between individuals and differences in their behavior are linked, this does not mean that they will behave in a prescribed way simply because they possess a particular gene.

If we do use the phrase "there is a gene for a particular behavior" what we really mean is that some part of the genetic material of the individual causes variation in the physical structure of that individual which in turn affects its behavior under some particular combination of environmental circumstances.

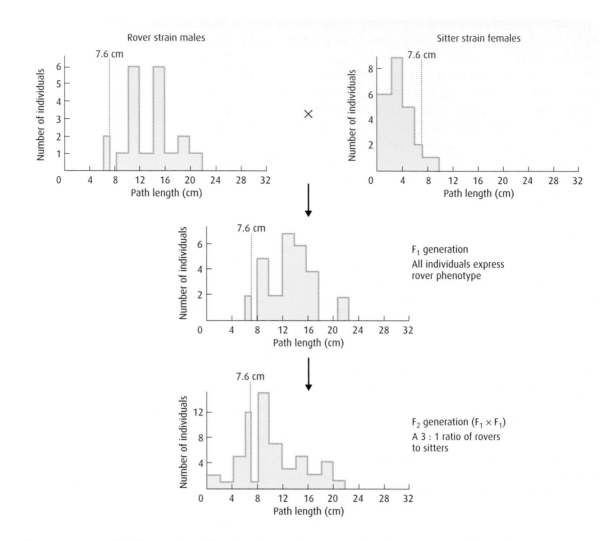

Fig. 4.1 Frequency distributions of path lengths of rover and sitter *D. melanogaster* strains and of F_1 and F_2 crosses involving them. (From DeBelle, J.S. & Sokolowski, M.B. (1987) Heredity of rover/sitter: alternative foraging strategies of *Drosophila melanogaster* larvae. *Heredity* **59**, 73–83. Reproduced with permission of Nature Publishing Group.)

to pinpoint the exact position of the gene to a site termed 24A3-C5 on chromosome number 2 of the *Drosophila* genome.

There are occasions, however, when a fly that has the homozygous *for*^{s} genotype will behave as a rover. This occurs specifically when the *for*^{s} fly finds itself in an environment that is short of food. Clearly under these conditions being a rover – moving to find food – is the only strategy likely to pay off. So why does the sitter genotype/phenotype persist? To answer this question Sokolowski and her colleagues reared large populations of *Drosophila* under two very different and carefully controlled sets of conditions. One group of flies were reared at very high larval densities and a second set were reared at very low larval densities. After 74 generations the proportions of sitters and rovers in each population was examined and in both cases strong directional selection was found to have occurred.

Under high density conditions the rover phenotype had become relatively more common, i.e. it had been selected for and the sitter phenotype had been selected against. At low population densities the opposite had occurred and an increase in sitter flies was recorded.

It seems likely that at higher densities individual larval flies would have to move further when they forage. The high density of competitors at the food source may make the distribution of the food itself patchy, or it might just be the case that a foraging fly needs to move a greater distance as it moves around and over its neighbors to find a place to feed. In this scenario if a sitter were to stay put it would probably starve and rover behavior would be selected for. But when the food is super abundant, and competition is low because larval densities are low, a sitter can do well and may do better than a rover simply because it doesn't pay the relatively high energetic costs associated with maggot movement. As we will see in Chapter 6 food in nature is patchily distributed and patches tend to vary in both quality and quantity. Population densities are also likely to vary from site to site and with time. So it should now be clear to you why the rover/sitter polymorphism persists.

So variation in just one of the 13,061 genes of *Drosophila* leads to an observable difference in behavior. Single gene effects have been recorded from a number of other species. For example, Walter Rothenbuler has demonstrated that the variation in behaviors of so called hygienic and nonhygienic honey bees, *Apis mellifera*, is related to variation in just two genes, each having two alleles and each affecting one behavior in a two-behavior sequence (Fig. 4.2).

Pleiotropic effects

It would be quite wrong to believe, however, that in all cases variation in a single gene is related to variation in a single phenotypic

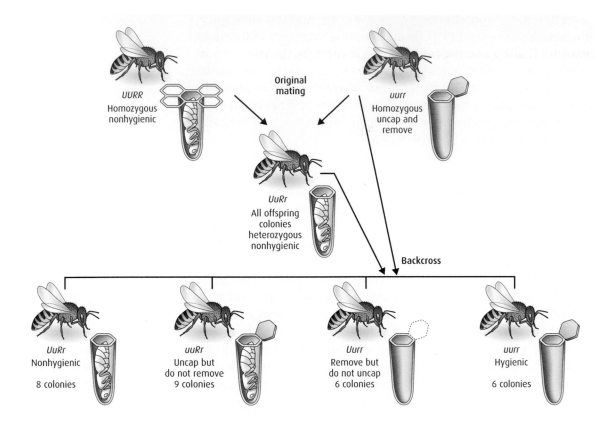

Fig. 4.2 A homozygous hygienic bee (uurr) will uncap (open) brood cells that contain a dead larva. It will then remove the corpse from the hive. Homozygous nonhygienic bees (UURR) perform neither of these behaviors. A cross between these two strains will result in an heterozygotous F_1 generation of nonhygienic individuals (UuRr) showing that the nonhygienic alleles are both dominant with respect to the hygienic alleles. The results of backcrosses produce a combination of behaviors. Some individuals behave in the typical hygienic or nonhygienic manner, but bees that uncap but then leave the corpse and bees that do not uncap, but will remove a corpse from an open cell, are also produced. This demonstrates that two genes are involved, an uncapping gene and a removal gene. (From Hall, M. & Halliday, T., eds (1998) *Behaviour and Evolution*. The Open University/Springer-Verlag, Berlin. Reproduced with permission of Springer-Verlag.)

character. Some genes are pleiotropic, by which I mean that a variation in one of them is expressed as a variation in a number of different phenotypic characters. For example in Chapter 3 we discussed the role of the *per* gene in the control of circadian rhythms in a wide range of species. The same gene, it turns out, has a role to play in *Drosophila* courtship. Male fruit flies "sing" to entice females to mate with them. The song is produced when the male extends and vibrates one of his wings and it consists of a hum and a series of pulses. The hum is thought to prime the female for mating and the pulses are thought to be the trigger that enables copulation to occur. Flies possessing different *per* mutations have songs that pulse at different frequencies. The adaptive significance of these variations in song, however, remains an area of speculation and debate.

Polygenic phenotypes

Other phenotypic variations are described as being polygenic because they result from the combined effect of variation at a number of loci. The courtship song of another insect, in this case the cricket, provides a good example of a polygenic behavioral trait. Like the *Drosophila* that we have just discussed, male crickets "sing" to attract females. Cricket songs are also mechanical in origin and they are produced when the animals open and close their specially modified forewing cases. When the wings open no noise is emitted, but as they close they rasp against one another and a sound pulse results. The members of each species sing a specific song characterized mainly by the duration of these pulses and the silences between them. Figure 4.3 depicts sonograms

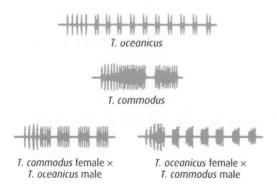

T. oceanicus

T. commodus

T. commodus female ×
T. oceanicus male

T. oceanicus female ×
T. commodus male

Fig. 4.3 A sonogram could be thought of as a "graph of sound" with time along the x-axis and the loudness of the noise on the y-axis. These pictures show the characteristic pulses of noise (chirps) that make up the songs of two cricket species of the genus *Teleogryllus*. Notice that the songs of hybrids have some of the features of each of the songs of their parents. (From Hall, M. & Halliday, T., eds (1998) *Behaviour and Evolution.* The Open University/Springer Verlag, Berlin; data from Bentley, D. & Hoy, R.R. (1972) Genetic control of the neuronal networks generating cricket song patterns. *Animal Behaviour,* **20**, 478–92.)

When we use the term **fitness** in the context of evolutionary or behavioral biology we are referring in some way to the reproductive success of an individual, to the number of copies of its genome that it manages to pass on to the next generation. I would strongly encourage you to think about individuals as vehicles for the propagation of genes by reading one or more of the many excellent books by Richard Dawkins.

The **classical fitness** of an individual can be thought of quite simply as the number of offspring it produces during its lifetime that go on to reproductive age themselves. But if you think about it this is really only one component of fitness. To use myself as an example I can achieve fitness by caring for my son William and doing all in my power to ensure that he survives to maturity, but I can also achieve an amount of fitness by doing the same for my brother's daughter Bethany. This is because through our relationship as uncle and niece Bethany and I share some of our genetic material in common. By caring for William I enhance my **direct fitness** but by caring for Bethany I enhance my **indirect fitness**. Clearly my **total** or **inclusive fitness** must take this into account and so must be a combination of both of these components. The table below shows the average coefficient of relatedness (r) between a father and son, an uncle and niece, and other relatives. These coefficients are simply the average probability that the two individuals concerned share a particular gene. As a cautionary note, however, it should be borne in mind that only those relatives that I actually help can be included in a calculation of my inclusive fitness, and that by acting to increase my indirect fitness I might in fact reduce my direct fitness. The calculation is not therefore a simple summation of the relevant coefficients.

Relationship	r
Parent/offspring	0.5
Full siblings	0.5
Grandparent/grandchild	0.25
Uncle/niece	0.25
Cousins	0.125

In terms of our understanding of animal behavior an appreciation of inclusive fitness is important. For example, it helps us to understand the evolution of helping behavior in a range of species. It also provides an explanation for the

that represent pictorially the songs of two cricket species and of hybrids between them. A comparison of these sonograms reveals that the hybrid songs have features of the songs of both of its parent species, but are distinct from them and from one another. This observation, together with the fact that song is a product of wing-case shape, musculature, and nervous system, suggests strongly that song production in these species is polygenic.

The environment and behavior

Hygienic bees revisited

Earlier in this chapter I mentioned in passing that despite their genotype sitter *Drosophila* behaved as rovers during times of food shortage. This is a clear example of the fact that genetic differences between individuals do not necessarily mean that those individuals are slavishly "required" to perform a particular behavior. Whilst the genetic component of behavior is undoubtedly important, genes have their effect against the backdrop of a complex environment and environmental considerations can strongly affect behavior.

The brood-cleaning behavior of hygienic honey bees provides a useful example of the interaction between genes, environment, and behavior. Remember that bees that are homozygous for both of the hygienic genes will uncap brood cells that contain a dead larva and remove the body, but unhygienic bees will not. Hygienic bees may have the genetic ability to perform their clean-up duties, but a feature of the division of labor amongst the workers in a hive is that there exists a particular age structuring in the performance of duties. Hygienic behavior tends to be performed most commonly by middle-aged bees, which will eventually switch behavior as they age, leave the hive, and become foragers.

evolution of the eusocial behavior of bees, ants, and wasps. These insects form colonies in which workers do not reproduce themselves, but instead contribute to the rearing of their siblings. These sterile workers have a coefficient of relatedness to their mother of 0.5, and if they were to breed would of course achieve $r = 0.5$ themselves. But by rearing their siblings they in fact gain an r of 0.75. This is a result of the sex-determination system of these insects. Female hymenopterans are diploid organisms having two copies of each chromosome in their cell nuclei. Males, on the other hand, are the product of an unfertilized egg and so possess only one set of chromosomes (they are haploids). Sisters that are the product of a single mating will therefore all inherit the same chromosome from their father and half of those of their mother, on average therefore they share 75% of their genome.

The switch to foraging occurs as a response to changes in the "internal environment" of the bee. Hive-bound workers typically have low concentrations of juvenile hormone in their blood. An increase in circulating juvenile hormone has been shown to trigger the onset of foraging. This process occurs naturally as the bee ages, but the actual timing of the switch can be varied according to the "needs" of the hive. For example, in a colony with a sufficient number of older foraging bees the younger bees will develop more slowly in behavioral terms.

Recently it has been shown that the expression of hygienic behavior in worker bees is also partly environmentally controlled. You will remember that to behave hygienically a worker bee must have the appropriate genotype. However, it would appear that for an individual bee both the amount of time it spends on its hygienic duties and the age at which it gives them up to become a forager are controlled in part not by the animal's own genotype and internal hormonal environment, but by the genotypes of its hive mates and therefore the genetic and social environment in which it lives.

To manipulate hive-level genetic make-up researchers first selected two behaviorally different strains of bees, one hygienic and one nonhygienic. This was achieved by first presenting a number of colonies with combs containing dead brood and then assessing the speed with which they cleaned them. Queens and drones were taken from the most hygienic colony and allowed to breed. The resultant colonies were again tested for their level of hygiene and drones from the most hygienic were mated to

Table 4.1 The expression of hygienic behavior in bees depends to some extent upon the hive environment. (Data from Arathi, H.S. & Spivak, M. (2001) Influence of colony genotypic composition on the performance of hygienic behaviour in the honey bee (*Apis mellifera* L.). *Animal Behaviour*, **62**, 57–66.)

Hive type	% hygienic bees behaving hygienically at any one time	Mean age of hygienic bees
25% hygienic	79	40 days
50% hygienic	40	19 days
100% hygienic	19	17.5 days
100% nonhygienic	0	–

queens from the previous generation. This was repeated for six generations and the result was a genetically homogeneous hygienic strain of bee. Exactly the same strategy was used to develop a comparison strain of nonhygienic bees.

Experimental hives were established using bees from the two strains, such that the proportions of hygienic and nonhygienic individuals in each of them varied. One hive contained only 25% hygienic bees, a second contained 50%, and a third 100%. A fourth hive was also established containing only nonhygienic individuals. At intervals a fixed area of freeze-killed brood was introduced to each of these hives and the levels of hygienic behavior and identities of the individuals performing it were recorded.

As the data in Table 4.1 show, the individual hygienic bees in a colony containing mostly nonhygienic bees are more likely to be engaged in cleaning duties at any one time than are their counterparts in hygienic hives. In addition, the harder working bees each cleaned a greater number of cells per day than did their less active counterparts. This is probably a function of the fact that there are numerically fewer of them to clean a comb of equivalent size. It is also apparent that hygienic bees in the 25% hygienic hive delay their switch to become foragers well past their usual middle age of around 19 days. So it would appear that in spite of their genetic similarity two individuals from the hygienic line might differ in the extent to which they perform hygienic behavior depending upon the environment in which they find themselves.

Learning, the modification of behavior

Some aspects of the behavior of an animal are performed in a rigidly predictable way such that each time a particular stimulus is presented a predictable response will be recorded. Presumably any variation in these behavior patterns that has existed in the past has been strongly selected against and eliminated by natural selection. But as I am sure you are well aware from your own experiences, not all behaviors can be characterized in this way. In many situations flexibility of response is a definite advantage, and so it should come as no surprise that many behaviors may be permanently modified as a result of experience. This modification can be attributed to learning.

Innate behavior

Most animals must perform a large number of quite varied behaviors that they have no opportunity to learn. This may be because they never meet their parents, or any members of the parental generation, to learn from, or because their survival depends upon them being able to perform the behavior very soon after they hatch or are born, allowing no time for learning. Such behaviors are termed **innate**.

The begging behavior of herring gull chicks that we discussed in Chapter 2 is an example of an innate behavior. You will no doubt remember that these chicks beg by pecking at the beaks of their parents in an attempt to stimulate them to regurgitate a meal. We also saw that a model bearing what would seem to be the most superficial resemblance to the beak of the parent bird would stimulate begging on the part of the chick. I am returning to this example because I want to make an important point about innate behaviors. Even though they can by their definition be performed with no practice, it would be wholly wrong to assume that they are inflexible in character, as Jack Hailman has demonstrated. Hailman presented herring gull chicks with models of the heads and beaks of either adult herring gulls or adult laughing gulls. The laughing gull (*Larus atricilla*) is a close relative of the herring gull and so the two models were very similar in shape. However, whereas the herring gull has a white head with a beak that is yellow with a red spot at its tip, the beak of a laughing gull is red all over and its head is black.

Initially the young birds showed no preference for either model, pecking at both whenever they were presented (Fig. 4.4). This is not a surprising finding given what we already know about the responses of young herring gulls to beak-like stimuli. But as the days passed a different pattern was revealed. With time the

Concept
Nature versus nurture

At an early juncture in the study of animal behavior two schools of thinking took up opposing positions on what came to be known as the nature–nurture debate. Psychologists became champions of the role of learning (nurture) in the development of behavior, whilst ethologists favored an investigation of the role of genes (nature). The debate was a fruitful one in that it stimulated a wealth of discussion and research.

Today, however, few would see the presumed dichotomy in these stark extremes, and the position that we should take is that genes and environment are both vitally important, as is their often complex interaction. After all without genes there would be no behavior, but without an environment there would be no behavior either.

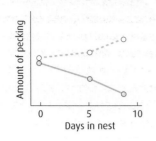

Fig. 4.4 As they age herring gull chicks learn to peck more at a herring gull model (dashed line) and less at a laughing gull model (solid line). (From Hailman, J.P. (1969) How an instinct is learned. *Scientific American*, **221**(6), 106.)

chicks began to show a strong preference for the herring gull model and became progressively less interested in the model laughing gull.

Of course, while these chicks may peck at both models their efforts will only ever be rewarded by their parents – who more closely resemble one model than they do the other. It would appear therefore that the chicks associate the herring gull model/stimulus with food and direct their efforts accordingly. So by a process that Hailman termed "perceptual sharpening" the innate behavior of these young birds is refined with time, or to put it another way – they learn.

Habituation

I remember once moving into a flat on a busy road junction while I was a graduate student in Edinburgh. At first I thought that the traffic noise was unbearable and I simply couldn't comprehend that I would be able to stay there for the full term of my lease. But of course a few weeks later when a friend new to the area visited me and complained about the noise, I found myself explaining that that I hardly noticed it. This is an example of the ability that we have to learn to ignore unimportant stimuli, a process termed **habituation**.

When should a crayfish tail-flip?

In Chapter 2 I described the process by which a stimulus detected by the sensory hairs on the abdomen of a crayfish can result in the animal performing a powerful tail-flip that projects it through the water at speed. You will recall that the purpose of this behavior is to enable the animal to escape from the source of the stimulation that could in many cases be a predator. Escape behaviors such as this one are prime candidates for habituation because not every stimulus is a predator.

If a crayfish is tapped on the abdomen repeatedly (about one tap per minute) the probability that a tail-flip will occur diminishes quickly. In fact after just 10 taps the response can diminish to zero – habituation has taken place and recovery from it can take several hours (Fig. 4.5).

But think for a moment about what must happen during a tail-flip. If the tail-flip can be induced when water movement displaces the sensory hairs on the abdomen surely we would expect a

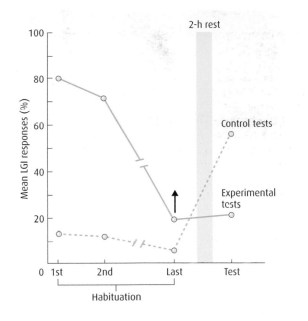

Fig. 4.5 Crayfish rapidly habituate to repeated tactile stimulation and recovery from it may take several hours (solid line). When the LGI is stimulated just before each tactile stimulation (broken line) no response occurs. This is a result of command-derived inhibition of the reflex at each trial. After a period of rest these animals show less habituation than those that had not had direct LGI stimulation. (From Carew, T.J. (2000) *Behavioural Neurobiology*. Sinauer Associates, Sunderland, MA; from Wine, J. (1975) Habituation and inhibition of the crayfish lateral giant fibre escape response. *Journal of Experimental Biology*, **62**, 771–82.)

single tail-flip to be sufficient to fully habituate the behavior because as the animal shoots through the water column those hairs must be repeatedly and vigorously stimulated. If this were to occur it could be a real problem for the individual. A stimulus that is repeated at 1-minute intervals is probably just a piece of floating vegetation rather than a real threat. But it might take more than one flip to outrun a persistent predator.

This problem is avoided because the LGI-controlled tail-flip reflex is its own protection against inhibition. Figure 4.5 illustrates this. If the LGI (lateral giant interneuron) is artificially electrically stimulated just before the tap stimulus is applied no behavioral response is recorded, and if the animal is tested 2 hours later (this time without a stimulation of the LGI) it will perform a tail-flip. This is evidence that the LGI can inhibit habituation. The mechanism by which this occurs is the same depression of synaptic transmission and command-derived inhibition that we discussed in Chapter 2.

Associative learning in *Aplysia*

The ability of an animal to learn by association is clearly biologically very important, and examples of associative learning involving

Focus on **conditioning**

Learning about the relationship between one stimulus and another (classical conditioning), and about the consequences of one's actions (operant conditioning) are biologically important in that they provide the individual with a level of predictive understanding of the environment in which it lives.

There can be few of you who are completely unaware of the work of the pioneering students of conditioning, Pavlov, Thorndike, and Skinner, but is worth revisiting the basics of their experiments.

The work of Pavlov relates to classical conditioning, the process by which an animal learns about the relationship between one stimulus and another. In his experiments Pavlov presented tethered dogs with food (the **unconditional stimulus**) and recorded in them a predictable **unconditional response** – they salivated. He then went on to present the unconditioned stimulus alongside a second and highly artificial **conditional stimulus** (a buzzer or a light). After just a few simultaneous presentations of the two stimuli it was found that the conditional stimulus alone was sufficient to elicit salivation (now termed the **conditional response**). To put it simply, the dog has learned to associate the novel stimulus with food.

Thorndike and Skinner both carried out experiments in which animals were housed in an unnatural environment and required to perform a novel behavior in order to escape or gain some reward (operant conditioning). Thorndike, for example, placed cats into a box that they could only escape from if they pressed a release lever. Initially they would do this quite by accident when scrabbling to get out. But as a result of this trial and error or **instrumental learning** they were very quickly able to affect an immediate escape when returned to the box. Similarly animals may be housed in a **Skinner Box** where they are required to peck a disc or pull a lever in order to release a food reward. Although the test subjects may initially need to be trained to do this by gluing food to the disc/lever, they too quickly learn to associate the performance of a novel behavior with a predictable outcome.

Link
Alarm calls habituate and dishabituate in a similar manner.
Chapter 5

all areas of the animal kingdom can be found in the literature. But the animal that has provided us with the most detailed understanding of the mechanisms involved in the process is not a lab rat or a pigeon or a primate, it is in fact a slug. The sea-slug *Aplysia californica* is a marine vegetarian that can grow to a massive 1 meter in length and reach a weight of up to 7 kilograms (Fig. 4.6).

In just the same way that Pavlov was able to train dogs to link a conditional stimulus (CS) with a conditional response (CR), Thomas Carew and his colleagues have been able to train *Aplysia* to associate one stimulus applied to the siphon with the subsequent application of another stimulus to the animal's tail. That is to say they demonstrated that *Aplysia* could remember that one stimulus predicted another.

In one of their experiments the researchers applied a mild tactile stimulus to the siphon of their test subjects and by doing so elicited the animals gill withdrawal reflex (a defensive behavior). If the stimulus is repeated at 90-second intervals the reflex rapidly habituates, a process that manifests itself as a sequential diminution in the intensity (amount and length of time) of the gill withdrawal (Fig. 4.7). Dishabituation, the reactivation of the reflex if you like, can be achieved if a sharp pinch is applied to the animal's tail. The fact that an habituated behavior can dishabituate makes biological sense, after all while it may be safe for *Aplysia* to "assume" that a repeated tapping on one part of its body is just a seaweed frond moving in the current, it certainly isn't safe to assume that a new stimulus is the same thing. In conditioning or associative learning terms we can think

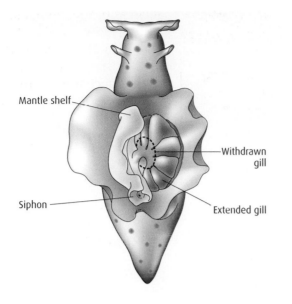

Fig. 4.6 When stimulated *Aplysia* withdraws its extensive gill to the position indicated. (From Kandel, E.R. (1979) Small systems of neurons. *Scientific American*, **241**(3), 67–76.)

of the periodic tactile stimulation of the animal as being an unconditional stimulus (US) and of the animal's gill withdrawal reflex as the appropriate unconditional response (UR).

To find out if *Aplysia* was capable of learning by association Carew and his colleagues paired a small electric shock to the tail (in this case the US) with a conditional stimulus (CS) that took the form of the tactile stimulation of the siphon. The treatment received by one group of animals involved the application of the US and then the CS immediately after it every 5 minutes (paired CS-US), whereas the stimulation applied to the animals in a second group involved an alternation of the US and CS every 2.5 minutes. After a training period of 30 trials the animals in the paired CS-US group exhibited a much stronger response to the US than did those animals in the unpaired US-CS group (Fig. 4.8). These findings suggest that not only is *Aplysia* able to learn that one stimulus predicts another, but that in some way information about the relationship between the two stimuli is retained in the memory of the animal for a period of about 4 days.

Aplysia memory

Experiments involving the neural pathways involved in the *Aplysia* gill withdrawal reflex have provided important insights into at least one of the mechanisms by which memory operates.

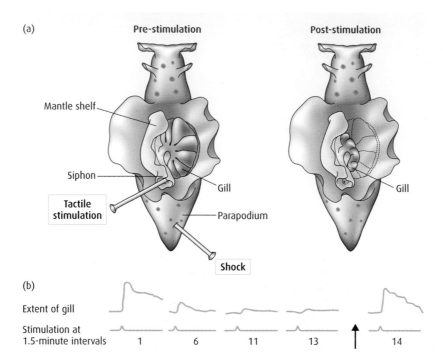

(a) Pre-stimulation   Post-stimulation

Mantle shelf

Siphon

Tactile stimulation

Gill

Gill

Parapodium

Shock

(b)

Extent of gill

Stimulation at
1.5-minute intervals   1   6   11   13   14

Fig. 4.7 Tactile stimulation of the siphon of *Aplysia* results in gill withdrawal (a). As (b) shows, the extent of the withdrawal lessens, the animal habituates, with progressive stimulations. Application of a small electrical stimulus to the animal's tail between stimulation 13 and 14 (arrow) results in a full gill withdrawal or dishabituation. (Reprinted with permission from Kandel, E.R. & Schwartz, J.H. (1982) Molecular biology of learning: modulation of transmitter release. *Science*, **218**, 433–43. Copyright (1982) American Association for the Advancement of Science.)

Look at the simplified neural pathway that is presented in Fig. 4.9. In this model three different sensory neurons each have the potential to stimulate a motor neuron that will in turn trigger the gill withdrawal reflex. One of these neurons provides a connection to the animal's tail, one links to the siphon, and the third connects with the mantle. Experimentally it is possible to artificially stimulate each of these neurons in order to simulate the kind of stimulation that we discussed when we considered the way in which *Aplysia* learns by association. By monitoring changes in the behavior of the individual neurons and the synapses that they form, we can determine their role in the learning process.

One way to do this is to "train" our neural model to associate one stimulus with another in exactly the same way that Carew and his colleagues did with the whole animal. So if stimulation of

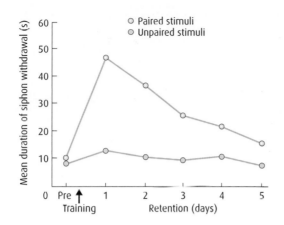

Fig. 4.8 After training *Aplysia* gill withdrawal is greatest for those animals in the paired CS-US group. It is also clear that the association is remembered by the animal for more than 4 days. (After Kandel, E.R. (1984) Steps towards a molecular grammar for learning: explorations into the nature of memory. In *Medicine, Science and Society*, ed. by Isselbacher, K.J., pp. 555–604, Wiley, New York, reproduced with permission of Elsevier; data from Carew, T.J., Walters, E.T. & Kandel, E.R. (1981) Classical conditioning in a simple withdrawal reflex in *Aplysia californica. Journal of Neuroscience*, **1**, 1426–37.)

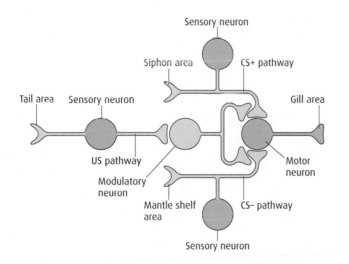

Fig. 4.9 A simplified diagram of the neural pathways that are involved in conditioning in *Aplysia*. (From Pearce, J.M. (1997) *Animal Learning and Cognition: An Introduction*. Psychology Press, Hove. Reproduced with permission of Thomson Publishing Services.)

the siphon area is always coupled with stimulation of the tail we have a paired CS-US condition and we would expect learning to take place. For comparison we would never follow mantle simulation with tail simulation and would expect no learning to occur.

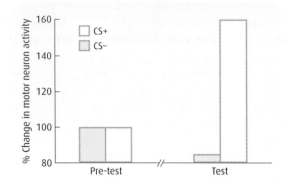

Fig. 4.10 The mean percentage change in motor neurone activity in *Aplysia* to stimulation of one sensory neuron is greater if it was paired with a tail shock (CS+) than if it was not (CS–). (From Pearce, J.M. (1997) *Animal Learning and Cognition: An Introduction.* Psychology Press, Hove, reproduced with permission of Thomson Publishing Services; data from Macphail, E.M. (1993) *The Neuroscience of Animal Intelligence.* Colombia University Press, New York.)

When such an experiment is carried out the results presented in Fig. 4.10 are obtained. They demonstrate that conditioning results in higher motor neuron activity in the US-CS condition. But how does this happen?

Stimulation of the tail causes the stimulation of the modulatory interneuron, which forms synapses with the sensory neurons from both the siphon and the mantle. This stimulation triggers the release of seratonin at these synapses. If one of the sensory neurons fires at the same time as this seratonin release occurs (as would happen in our CS-US condition), a relatively permanent change in the sensory neuron synapse occurs with the result that it will subsequently release higher than normal amounts of neurotransmitter. When this happens the response of the motor neuron is amplified and the strength of the reflex will increase. So in this example learning is brought about via an increase in the effectiveness of the neural pathway.

Social learning

From our own day-to-day experience it would seem obvious to us that individuals should be able to learn from one another. For example, how often do you try something new because you have watched a friend try the same thing and clearly enjoy it? How often have you learned at least the rudiments of a task by first watching it performed? I'm sure that this will be a common occurrence during your time as a student.

Animals may learn from one another by imitation or by mimicry. The distinction here is a subtle but important one. Learning by imitation involves the performance of a novel behavior pattern that will result in an immediate reward. A behavior that is learned by mimicry, on the other hand, will have no tangible immediate reward associated with it. Young song birds, for example, learn some of the specific detail of their song from adult tutors. The rewards associated with this learning will not be available until the birds sing as adults and so learning by mimicry does clearly happen.

Day-old chicks peck instinctively at seeds and seed-like objects. In experiments involving an actor chick that was allowed to peck at a bead coated with the bitter-tasting substance methylanthranilate in the presence of an observer chick that was itself unable to peck, researchers have found that the observer would subsequently avoid pecking similar-looking beads. The conclusion drawn, i.e. that the observer learned, by imitation, to avoid methylanthranilate because it had seen the reaction of the actor to the bitter taste, is further supported by the fact that if the actor's bead is replaced with one that does not have a bitter taste the observer will not subsequently avoid it.

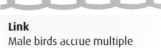

Link
Male birds accrue multiple benefits by singing.
Chapter 5

Application Learning to be wild again

Animal reintroductions and translocations are potentially important strategies in the conservation of endangered species. Unfortunately, however, most attempts to carry them out are unsuccessful. These failures can be attributed to a wide range of causes, but chief amongst them are behavioral failings on the part of the animals involved. As a result of their period of captivity or isolation populations and individuals may simply lack the skills that they need to survive in their new home. But all is not lost. Once these problems have been identified conservationists are able to use the principles of animal learning to train and prepare animals for release.

One strategy is to provide captive animals with an environment that mimics the one into which they will be released as closely as possible. For example, the first releases of captive-bred golden lion tamarins (*Leontopithecus rosalia*) back into their native Brazilian forest were unsuccessful. Observations of the released animals showed that they were woefully inadequate when it came to moving around in a complex forest setting. As a response to this the keepers responsible for tamarin production increased the complexity of the enclosures in which the animals were reared – affording them more opportunities to develop their locomotory skills. In addition prerelease enclosures were established in the zoo grounds. These wooded areas presented still more realistic opportunities for locomotion and navigation and also provided some opportunity for the animals to develop more natural foraging skills. Releases involving these trained animals have met with more success.

In some cases however one specific skill is required rather than a generally improved ability to use the environment. Very often animals that live in a predator-free environment (as a result of captivity or isolation) have poor or nonexistent antipredator responses. When valuable individuals are released or translocated to found new wild populations as part of a conservation exercise, the last thing that we would want is for them to be killed and eaten. The following examples illustrate two of the training strategies that are currently used to prepare such animals for release.

Animals can learn that a stimulus is dangerous if they observe a member of their own species acting in a fearful way, they learn by imitation. Eberhard Curio illustrated this principle very well when he simultaneously exposed two blackbirds (*Turdus merula*) that could hear but not see one another to two different stimuli. One of them was shown a stuffed owl (a potential predator) to which it responded appropriately with mobbing calls. The other bird could hear the mobbing calls but the model that it had been shown was a harmless honey-guide, an unfamiliar species (from another continent). The blackbird mobbed the honey-guide, and continued to do so whenever it encountered it again.

The other way that animals can be trained to respond appropriately to a predator is to teach them to associate its appearance with an unpleasant experience – to condition them. This is exactly the strategy that Andrea Griffin and her colleagues have adopted in their experiments that have been designed to demonstrate that tammar wallabies (*Macropus eugenii*) can be prepared for reintroduction into those areas of mainland Australia where they have recently become extinct.

During training one group of wallabies was presented with the paired stimuli of a stuffed fox (a novel predator, but one that reintroduced animals are likely to meet) and a simulated attempt by a person with a net to capture them. (Capture attempts by a person with a net consistently produce alarm behaviors in tammar wallabies and so this was deemed to be an appropriate aversive stimulus.) A second (control) group of wallabies were presented with the same two stimuli, but they were not paired.

After training, animals from both groups were shown the stuffed fox and their reactions were monitored. As Fig. 4.11 shows the control animals were less vigilant and more relaxed than the conditioned subjects, suggesting that the latter had learned to fear foxes.

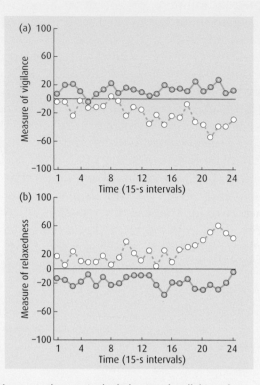

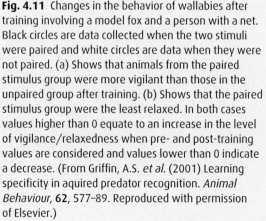

Fig. 4.11 Changes in the behavior of wallabies after training involving a model fox and a person with a net. Black circles are data collected when the two stimuli were paired and white circles are data when they were not paired. (a) Shows that animals from the paired stimulus group were more vigilant than those in the unpaired group after training. (b) Shows that the paired stimulus group were the least relaxed. In both cases values higher than 0 equate to an increase in the level of vigilance/relaxedness when pre- and post-training values are considered and values lower than 0 indicate a decrease. (From Griffin, A.S. *et al.* (2001) Learning specificity in aquired predator recognition. *Animal Behaviour*, **62**, 577–89. Reproduced with permission of Elsevier.)

Navigation

In an ethological experiment that is now regarded as a classic of its kind, Niko Tinbergen demonstrated that the female digger wasp (*Philanthus triangulum*) is able to return directly to her tiny

When they emerge from their nests hatchling turtles make a dash for the sea. It is important that they get there as soon as possible because if they are still on the beach when the sun rises they will be easy pickings for a range of predators and also vulnerable to dehydration. We have known for some time how they orientate towards the sea. Young turtles exhibit a strong positive phototaxis – they are drawn towards sources of bright light. On a beach the brightest natural horizon from the perspective of a hatchling is always seaward. But after millions of years of successful sea finding turtles today face a real problem – anthropogenic photopollution. The brightest sources of light of a beach today are unlikely to be natural ones. They are more likely to be an urban glow, road lighting, or an abandoned beach fire. Photopollution results in disorientation and a disruption of the turtle's sea-finding ability. Large numbers of hatchlings are commonly lured onto roads where they are crushed, or burned to death in abandoned fires, or become lost and exhausted in the dunes behind the beach. In Florida alone it is estimated that approximately one million hatchling turtles succumb to the lure of non-natural light sources each year.

What can be done to ameliorate this situation? By carrying out behavioral studies involving choice tests Blair Witherington has established that turtles respond more strongly to some wavelengths of light than to others (Fig. 4.12). Most of the species that he tested showed some aversion to more yellow light but strong attraction to blue light. So one practical step that could be taken would be to ensure that lights close to beaches are yellow.

Other steps that could be taken involve shielding the turtles from light sources. Increased use of directional lighting angled away from beaches might help in this respect. In the case of the glow from a distant town effective dune management to maintain dune height and dune-top vegetation can be highly effective. Turtles are also attracted more strongly to a constant light

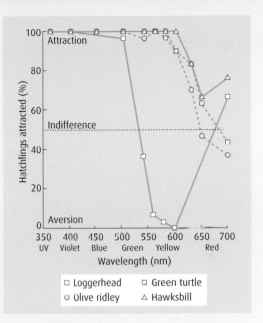

Fig. 4.12 Turtles show a stronger level of attraction to low wavelength light, and a stronger aversion to light with a longer wavelengths. (From Witherington 1997. Reproduced with permission of Cambridge University Press.)

source than they are to an intermittent one. To help in this respect conventional lighting along paths to turtle nest beaches in Australia has been replaced with flashing lights.

Witherington, B.E. (1997) The problem of photopollution for sea turtles and other nocturnal animals. In *Behavioral Approaches to Conservation in the Wild*, ed. by Clemmons, J.R. & Buchholz, R., pp. 302–28. Cambridge: Cambridge University Press.

nest burrow after a provisioning flight because she first memorizes the relative positions of landmark objects in her environment. In his experiment Tinbergen surrounded a wasp nest with a ring of pine cones. Then after the wasp had emerged from the burrow and flown away, he moved the ring of cones a small distance so that the nest was now outside of it. On her return the wasp flew to the center of the ring and not to the burrow. I am sure that you are aware of a similar use of geographical references as you navigate your own environment. But I'm equally sure that you are able to navigate a familiar environment (the interior of your house for example) in darkness when no landmarks are available, suggesting that landmark memory alone is not a sufficient explanation of the ability of an animal to find its way.

Trail laying

When he sallied forth to do battle with the Minotaur the hero Theseus was advised by Ariadne to unwind a trail of thread as he made his way into the labyrinth. It was Ariadne's hope that having slain the beast Theseus would be able to follow this thread and return to her in safety. This of course he did, although the rest of the story didn't quite play out the way that Ariadne wanted it to. Hansel and Gretle tried a similar tactic when they were led into the forest to be abandoned. But their trail consisted of pieces of bread and of course the birds ate it – a pity they didn't have a ball of thread.

Trail laying and trail following as a navigational method are common throughout the animal kingdom. Ants, for example, use pheromone trails as a method by which a number of foragers can efficiently exploit a newly discovered food source. When it finds a food source that is too large for it to exploit successfully alone a foraging ant will return quickly, and by a very direct route, to its nest (Fig. 4.13). As it does so it deposits a pheromone trail on the ground behind it. At the nest the returning individual performs stereotyped behaviors designed to recruit others to the food source. By following the trail these recruits are able to go directly to it. As each of them returns to the nest they too deposit pheromones and so the trail is reinforced. Eventually the food source will become exhausted and animals will stop returning from it. No trail reinforcement will take place and quite quickly the trail will disappear. Their short-lived nature makes pheromone trails a particularly suitable navigational aid in this situation. If they

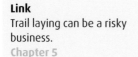

Link
Trail laying can be a risky business.
Chapter 5

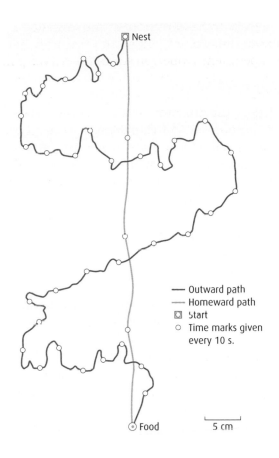

○ Nest

— Outward path
— Homeward path
▣ Start
○ Time marks given
every 10 s.

⊙ Food   5 cm

Fig. 4.13 Foraging and return path of the ant *Tetramorium caespitum*. The two lines mark the outward and homeward paths, respectively. Nest, start of foraging trip; white circles, time marks given every 10 s to and from the food source. N, nest; F, food site. (From Shen, J.X. *et al.* (1998) Direct homing behaviour in the ant *Tetramorium caespitum* (Formicidae, Myricinae). *Animal Behaviour*, **55**, 1443–50. Reproduced with permission of Elsevier.)

were too long lasting many ants would waste valuable foraging time following trails that provided no reward.

Dead reckoning

Look again at Fig. 4.13. It shows that whilst the outward route taken by the forager is a particularly tortuous one taking a considerable period of time, the return journey was remarkably direct and very rapid. Obviously as the discoverer of the food this individual did not have a trail to follow to the nest. So just how did it find its way home?

Navigation of this type is variously referred to as path integration or dead reckoning. The latter is a corruption of the term deduced reckoning and refers to an individual's ability to deduce its current position in relation to another location by taking into account the direction(s) and distance that it has traveled between the two.

Deducing direction and distance

In order to deduce the direction in which it must travel to return to an earlier location an individual must measure the various direction changes that it has made during its journey and from them compute an appropriate return vector. In order to do this some species such as our own are able to make use of internal cues. For example, the sensory cells of the semicircular canals in your ear, part of your vestibular system, provide you with information about the rotation of your body through space. Without your realizing it you can use this information to navigate a dead-reckoned path through a familiar space in darkness.

Ants, in common with a number of other species, on the other hand, rely upon external cues to guide them to their ultimate destination. The desert ant *Cataglyphis fortis* is capable of the kind of direct homing behavior illustrated in Fig. 4.13. If we were to allow a desert ant to find a food source, but then before it had a chance to set off on its return journey we moved it 2 meters to the east of the food and then let it go, the path that is followed would be direct but it would end at a point approximately 2 meters to the east of the nest. The angle of travel and distance walked would both be right but the end point would be wrong. Evidently the ant was unable to take into account our experimental displacement of it and to thereby correct its trajectory.

Findings such as these suggest that desert ants use global environmental cues rather than local landmarks by which to navigate. In fact desert ants use the position of the sun, or more precisely the angle between their direction of movement and the position of the sun relative to the horizontal plane (the solar azimuth). This means of course that they must be able to take into account the movement of the sun across the sky as the day progresses. Evidence that they can do this comes from a remarkably simple experiment in which a homing ant is trapped mid-journey under a lightproof box. When it is released some hours later the sun has of course changed position. The ant however will continue along its original trajectory, seemingly oblivious to our intervention.

It would be wrong to think that ants ignore landmarks when navigating. Research carried out by Tsukasa Fukushi of the Miyagi University, Japan, has demonstrated that wood ants *Formica japonica* use visual landmarks for navigation in preference to chemical trails or to celestial cues. When carrying out a displacement experiment similar in essence to the one that I have described above, Fukushi found that rather than behaving like desert ants and following a parallel path to the correct one, the pathways of ants displaced to the east or west of their capture position followed paths that looked initially as if they would lead the ant home. But on closer inspection it turned out that these paths all had a trajectory that would cause them to converge at a point some 13 meters behind the nest site (Fig. 4.14). Fukushi hypothesized that the convergence occurred because the ants were targeting a landmark somewhere in the distance and using it to guide them home. To test his hypothesis he carried out a series of experiments using large sheets to block out various components of the visual field of the ants; in some cases the skyline was obscured and in others the vegetation features closer to the ground were removed from view. The results of some of these experiments are reproduced in a simplified form in Fig. 4.14. These findings demonstrate that the ants were in fact using features of the skyline, prominent tree-tops for example as a navigational guide.

It has been demonstrated that ants will regularly stop, turn, and stare at a prominent landmark feature as they move away from it during a foraging trip. It is thought that during these **learning walks** as they are known, ants commit to memory key objects in their visual field and that on subsequent trips they can compare these snapshot memories with actual views. In this way thay are able to deduce information about the distance and direction of their goal.

Paul Graham and Thomas Collett of the University of Sussex in the UK have demonstrated that wood ants of the species *Formica rufa* use this kind of visual information to compute distance during their foraging trips. They trained ants to follow a 1-meter-long path to food that involved them walking parallel to a 20-centimeter-high wall at a distance of 20 centimeters from its base. If the ants were using the skyline created by the top of this wall as a landmark feature by which to navigate, a simple way in which they could do it would be to keep the top edge of the wall at an appropriate position on the retinal field, in this case represented by an elevation of 45°.

To test the hypothesis that the ants can judge distance in this way the team carried out experiments in which either the height

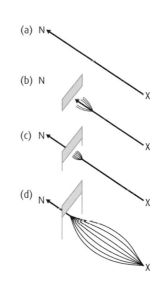

Fig. 4.14 Average path directions of ants moving from a release point (X) to their nest (N) when: (a) they have an unobstructed view of the vegetation beyond the nest; (b) a solid barrier is placed on the ground (view of distant vegetation is obstructed as animals approach); (c) as (b) but barrier 10 cm from ground to allow ants to pass under (view of vegetation restored when the ants clear the barrier); (d) an elevated barrier obstructs the view of the upper portion of the vegetation from the outset (the view is restored when the ants pass under the barrier). (Adapted from data in Fukushi, T. (2001) Homing in wood ants, *Fornica japonica*, use of a skyline panorama. *Journal of Experimental Biology*, **204**, 2063–72.)

Graham, P. & Collett, T. (2002)
View-based navigation in
insects: how wood ants
(*Formica rufa* L.) look at and
are guided by extended
landmarks. *Journal of
Experimental Biology*, **205**,
2499–509.

of the wall, or the distance from the wall base that the ant started its journey, was altered. In either case the effect would be a mismatch between the "correct" 45° elevation and the actual elevation of the top of the wall. So for example some ants were placed 20 centimeters from a 40-centimeter-high wall. In this scenario the elevation of the top of the wall is of course too high, and as would be predicted the ants were observed to move gradually away from the wall as they walked, thereby lowering the elevation. Similarly ants placed 40 centimeters away from a 20-centimeter-high wall (elevation too low) moved towards the wall as they walked. So it would appear that in both cases the ants could use the mismatch between the elevation that they had learned and the one that they observed to correct their course.

Mammals and birds have a number of cues that they can use to estimate their distance from a landmark. In addition to the retinal image properties that we have shown ants to use, the visual systems of the members of these taxa should enable them to take into account factors such as binocular convergence and disparity, motion parallax, and accommodation (the automatic focusing of the eye). However, surprisingly little is known about the means by which this information is actually used.

In addition to their work on wood ants, Thomas Collett and his coworkers have shown that gerbils can make use of retinal image size to judge distance. To do this they trained gerbils to travel towards a feeder at a fixed distance in front of a black cylinder (the only prominent landmark in their environment). The gerbils traveled closer to its base before beginning to search for food when the cylinder was replaced with a smaller one. This suggests that they were matching the retinal image size that they had learned with that which they observed. But when the cylinder was replaced with a larger one the gerbils looked for food at roughly the same distance from its base as they did during training; this shows that other cues were also utilized. If humans are subjected to tests that are broadly similar to those used by Collett to investigate the use of retinal image size in distance estimation in gerbils, similar results are obtained. But if pigeons are tested there is little if any evidence to suggest that they use the same cues.

Using spatial information from multiple landmarks

In reality of course it is highly unlikely that an environment will contain a single landmark. So the possibility must exist for animals to use information derived from an array of several landmarks simulataneously. Marcia Spetch and her coworkers at the University of Alberta, Canada, have carried out a number of experiments designed to increase our understanding of the ways in which a range of species might use this kind of information.

In one set of experiments individuals were trained to search for a goal (a position on a touch screen) that had fixed coordinates with respect to an array of four on-screen objects, and was the same distance from each of them (Fig. 4.15). During tests the relative positions of these objects were altered and their distances from the center of the array were increased. The results show that humans use information about the configuration of the various objects in the array to both identify the positions of the four landmarks and to identify the search area. By doing so they correctly search for their goal at a position equidistant from the location of each of the objects in the array without taking into account the identities of those objects. Pigeons, on the other hand, do not appear to use the configurational information in the same way. As Fig. 4.15 shows, they do search an area that is the right direction from the position of one of the objects (in this case the one in top left position), and so they do appear to use configurational information rather than object identity to solve this component of the problem. But to solve the second component of the problem – where to actually search – they do not. They fail to take into account the expansion of the array and appear to have learned that the goal is a fixed distance from one component of the array rather than equidistant from each of them. In similar tests gerbils use the information provided by their environment in a manner more similar to birds than to humans. Honey bees, on the other hand, do initially appear to solve such landmark-finding problems and so at first impression might be thought to be processing spatial information in a manner similar to humans, but closer inspection of the data suggests that the insects do not in fact use configurational information. Rather it would appear that they treat each object in the array as being independent from the others.

Cognitive maps

The ability of an animal to define various locations within its environment and then integrate this information is the basis of the cognitive map. Animals that have a cognitive map are able to

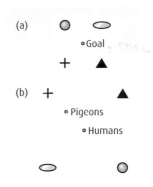

Fig. 4.15 The performance of humans and pigeons during expansion array testing. (a) Training situation. (b) Expansion test. (From Cheng, K. & Spetch, M.L. (1998) Mechanisms of landmark use in mammals and birds. In *Spatial Representation in Animals*, ed. by Healy, S., pp. 1–17. Oxford University Press, Oxford.)

Racing and homing pigeons are regularly taken by their owners to release sites several hundred miles from their home lofts. Amazingly the vast majority of these birds are able to quickly return home without any prior experience of the journey that they undertake. A good bird can home a distance of 1200 miles in just 2 days given the right weather conditions. How these feats of navigation are accomplished has been a key focus of animal behavior research.

For released birds how to find home is a two-stage problem. Firstly they need to determine their current position and their position relative to home, for this a map is needed. Then having determined their destination they need to know which direction to fly in and how far to fly, for this they need a compass and a means of measuring distance.

The compass

Based on the available evidence it seems highly likely that the position of the sun is used by pigeons as a compass. It seems that when they are released the birds are able to relate the position of the sun in the sky to their internal body clock and, by using that information, find the compass bearing that will take them home. Evidence in support of this idea comes from birds that have had their body clocks disrupted. For example, if a bird's clock is shifted forwards by 6 hours so that at 9 a.m. in solar terms the bird "thinks" that it is 3 p.m., any inferences that it makes about the position of the sun will be incorrect by some 90°. In such cases birds that should fly due south upon release tend to fly east.

This does not mean that pigeons are unable to home on a cloudy day or at night, they often do. So it would seem that they are able to derive compass information from a variety of other directional cues. For example, magnetic storms adversely affect homing birds and attaching a magnet to a pigeon on a cloudy day is sufficient to disrupt its homing ability. This treatment is designed to disrupt the bird's ability to "read" the earth's magnetic field and these observations suggest that pigeons do use a magnetic sense to orientate by. How exactly this is achieved remains the subject of debate, but two promising theories have come to the fore. One involves the idea that magnetite molecules in some tissues act as tiny compass needles, the other suggests that individuals are able to sense changes in biochemical reaction rates caused by differing magnetic fields. Homing is not however disrupted by a magnet on a sunny day, suggesting that the sun compass is probably the main orientation tool used.

The map

Birds that are clock shifted in the manner described above and then released at a site with which they are familiar make less of an error in terms of the direction of flight that would be predicted if the only information that they utilized was that derived from their compass sense. These birds are able to take into account information from a range of other inputs, which considered together must form the map that the animal uses to navigate. There is evidence that pigeons are able to make use of visual features in the landscape to guide them home. This is particularly the case in the vicinity of the loft, an area with which they are intimately familiar. But visual landmarks on a larger scale are used too.

Some evidence points to patterns of infrasound as a potential map. Pigeons can hear these ultra-low-frequency cues that radiate from steep-sided topographic features and are generated continuously by interfering oceanic waves. At least three major pigeon races have been disrupted under strange circumstances recently, and Jonathon Hagstrum of the US Geological Survey has suggested disruption of the birds' infrasound map by the sonic booms of passing Concorde planes is a potential explanation.

However the most convincing evidence for the basis of the pigeon map relates to the birds' sense of smell. Pigeons that have been smell-blinded by either severing the olfactory nerve or temporarily anesthetizing the olfactory receptors are unable to navigate home over a large distance. These birds can home over small distances, presumably because in this situation they use local landmark navigation. Similarly birds that have been raised in a loft that has been fitted with baffles to deflect the flow of air into it by 90°, such that air from the north enters the loft from the east, etc., learn an inappropriate olfactory map and are subsequently unable to navigate correctly.

plot routes through their surroundings based upon the information that they have stored in the map. For example, rats can be trained to swim through opaque water to the safety of a platform that is invisible to them because its surface is fractionally below that of the water. In an experiment involving one of these water mazes the rat is always placed into the water at the same location during training and it will quickly learn to swim a very direct route to the platform. Then during testing one of two parameters of the test are varied. In some cases the rat is placed into the water at a novel location; in others the rat is placed at the original location, but the platform is moved. Figure 4.16 provides stylized, but none the less typical, results from such an experimental procedure. When the rat is moved but the platform remains in the same place the route taken is almost as direct as during training. In these trials the rats must have used their cognitive map to deduce the position of the platform. But look what happens

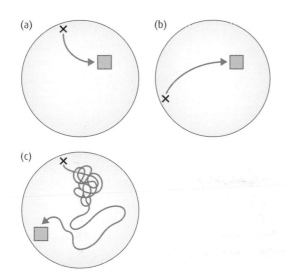

Fig. 4.16 A rat can be trained to swim from a point (X) to a submerged platform along a direct route (a). If the rat is placed into the water at a novel position (b), it will still swim directly to the platform. If the platform is moved (c) the rat will search the area in which it had been trained to expect the platform, before eventually locating its new position. (Adapted from Morris, R.G.M. (1981) Spatial localization does not require the presence of local cue. *Learning and Motivation*, **12**, 239–60. Reproduced with permission of Elsevier.)

The results of a large number of studies of the vertebrate hippocampus suggest that this area of the brain is linked to the ability of the individual to utilize spatial information. Neurobiological studies suggest that specific areas or cells of the organ provide very specific information about the location of an individual. Lesion studies reveal that damage to the hippocampus results in impaired spatial ability. And comparisons of animals with very different spatial information needs reveal that they have differently developed hippocampi.

Hippocampal place cells

Information collected via electrodes from individual cells within the hippocampus of free-moving rats has revealed a class of cells that have been dubbed **place cells**. Experimenters have discovered that each of these cells is sporadically active as the rat navigates its environment. However their activity is far from random. Each cell fires maximally when the rat reaches a particular location. The cells do not appear to respond to a location in space, but to particular landmarks, or combinations of landmarks that have a particular spatial relationship to the current position of the animal.

Hippocampal lesion disrupts navigation

Earlier in this chapter I described the ability of a trained rat to make use of the information provided by its cognitive map to solve a water maze. Rats that have been subjected to hippocamal damage lose the ability to carry out this task.

Hippocampal lesions performed on homing pigeons also result in disrupted spatial ability. Lesioned birds released more than 30 kilometers from their home loft were unable to home even if they were familiar with the release site. But they did at least set off in the right direction, so the hippocampus does not seem to be involved in their compass sense. If these birds were confined to their home loft for a sufficient period after their operation they eventually regained the ability to home to it, but none of them was able to learn to home to a new loft. Results such as these suggest that in pigeons the hippocampus is involved in both the acquisition and storage/retrieval of spatial information.

Comparative studies

Further evidence of the importance of the hippocampus in spatial memory comes from comparative studies of passerine bird species.

when the platform position is changed. The rats do eventually find it, but only after they have carried out an exhaustive search of the area in which they expected to find it based upon their map.

Migration

As feats of navigation the long-distance homing behaviors of pigeons are certainly impressive. But in my opinion the most amazing animal movement behaviors are the periodic mass migrations undertaken by a wide range of species. Migrations can be defined as persistent and very direct movements taking individuals from one locale to another clearly defined location, and often to a different habitat. Along the way other behavioral motivations (breeding for example, and perhaps even feeding) are often suspended. A key feature of migratory movements is that they have very distinct departing and arriving behaviors. Animals migrate because it benefits them to do so. For example, it may allow them to escape the rigors of winter or enable them to reproduce in a safe and/or particularly productive environment. But there are costs to migration too. The often long journeys are arduous in themselves, and in today's quickly changing world reliance on not one but two habitats and possibly staging areas between them places a number of species in a position of considerable conservation concern.

In Chapter 3 I mentioned that the onset of migration is under endogenous control involving the combined action of environmental cues and internal rhythms. In this chapter I want to briefly consider the journey itself, and in particular the ability of individuals to get to the right place.

Green turtles *Chelonia mydas*, sockeye salmon *Onchorynchus nerka* and blackcap warblers *Sylvia atricapilla* may seem to have little in common. But when it comes to migration they do face a similar problem. How do they navigate through thousands of kilometers of unfamiliar terrain and then arrive at a particular destination, and in the case of some returning individuals to within meters of a site that they have used in the past? Try for a moment to imagine though how challenging it must be to carry out research in this area. The turtle journey involves the animals shuttling back and forth over the 2500 kilometers of open Atlantic Ocean between the coast of Brazil and Ascension Island. Thousands of salmon fry must leave the breeding rivers for every fish that eventually return, making it almost impossible to follow an individual's progress. And any laboratory-based research involving these animals must also take into account the physiological transition made by the animals as they move from freshwater to saltwater and back again. But despite these difficulties some impressive advances in our understanding of migration have been made.

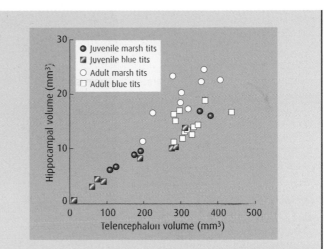

Fig. 4.17 Development of the hippocampus in storing (marsh tits) and nonstoring species (blue tits). (From Healey *et al.* 1994. Reproduced with permission of Elsevier.)

Marsh tits (*Parus palustris*) are avid food hoarders. An individual bird might store 50–100 seeds in a single morning, and across the course of a winter it will store or cache and successfully retrieve literally thousands of items of food. These items are cached singly, often meters apart, and may not be retrieved for several days. A number of experiments have been carried out to demonstrate that retrieval depends upon memory and not upon cues from the seeds themselves. In fact these birds can remember where a food item was hidden, what kind of food it was, and whether or not the cache has already been used. The closely related blue tit (*Parus caeruleus*) does not hoard food, and we would therefore assume that it does not have the same spatial memory needs.

Figure 4.17 shows that although juvenile marsh tits and blue tits have similar hippocampal volumes (none of these individuals will have stored seeds), the hippocampi of adult marsh tits are far more developed than are those of blue tits. These results illustrate the point that a large hippocampus is related to enhanced spatial memory, but they also suggest that hippocampal enlargement does not happen until the bird has experienced food-storing behavior. This suggestion has been verified experimentally.

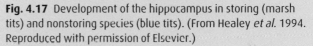

Healy, S.D., Clayton, N.S. & Krebs, J.R. (1994) Development of hippocampal specialisation in two species of tit (*Parus* spp.). *Behavioral Brain Research*, 81, 23–8.

Plate 4.1 Turtles, like the one resting here, undertake long migratory journeys. © C. Waller.

In terms of the ability of individuals to navigate over long distances it seems likely that animals use the same cues during migratory movements as they do during other forms of navigation. It is well documented that the sun, stars, and polarized light fields are used as compasses by birds. And salmon are known to use their sense of smell to home in on their spawning grounds. When it comes to finding a very specific location at one or other end of the migratory journey, a territory that had been used in the previous year, for example local landmark recognition, must be crucial.

Magnetic fields: maps or compasses?

Both migratory birds and turtles are able to sense the earth's magnetic fields and may be able to use them in navigation. Juvenile turtles are certainly sensitive to changing magnetic fields in the laboratory, but recent experiments that involved strapping magnets to the backs of migrating adult turtles and then monitoring their progress via satellite tracking have failed to show the expected disruption in their migratory ability. In this case then it would seem that either the magnetic sense is not important, or that the animals are able to use information from a variety of different sources.

The annual migrations of Australian silvereyes *Zosterops l. lateralis* between their Tasmanian breeding grounds and their Australian nonbreeding quarters can however be affected if the bird's

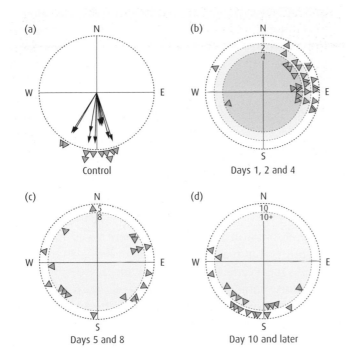

Fig. 4.18 Orientation behavior of silvereyes before and after treatment with a short, strong magnetic pulse. (a) The control orientation before the pulse; the arrows represent the mean vectors of the ten birds based on eight or nine recordings each; the triangles at the periphery of the circle indicate the respective mean directions. (b–d) The triangles represent the headings of the ten birds on the days shown (day 1 is the day of treatment).

magnetic sense is disrupted. When Wolfgang Wiltschko and his colleagues subjected migrating birds to a strong magnetic pulse they found that the orientation behavior of their subjects differed significantly to that of their untreated control birds (Fig. 4.18). From these results the conclusion may be made that adult silvereyes rely upon magnetic field information during their migratory journeys. I say adult birds here because earlier experiments using juvenile birds did not produce the same results. It would appear that the age of the subject animal is linked to its orientation behavior.

With age one assumes comes experience. Adult migrants have made the journey before whereas juveniles have not. This difference in experience has led Wiltschko and his team to suggest that in this case magnetic field information is used as a map rather than as a compass. The logic of this suggestion is as follows. The adult test birds and the adult control birds are familiar with the range of navigational factors in both their nonbreeding and breeding areas and along the migratory route between them. They have therefore had ample opportunity to fully develop a navigational map, which is assumed to include in it some information about magnetic gradients in their environment. The juveniles, on the other hand, have had little experience of their fledging area and no experience of the migratory route or destination. As a consequence they would not be expected to have developed a map and must instead be relying on innate information to determine their migratory direction. The fact that juveniles were unaffected by the magnetic pulse suggests that it does not affect compass sense. The effect that it has on the adult birds must therefore be affecting their map sense.

The genetics of migration

The previous discussion suggests that there are innate components to migratory navigation. Further evidence that this is certainly the case has come from the very impressive body of work on the migratory behavior of European blackcap warbler (*Sylvia atricapilla*) populations carried out by Peter Berthold and Andreas Helbig. Discrete breeding populations of these warblers migrate to discrete wintering ground. Hungarian blackcaps, for example, migrate through the Balkans and Turkey to winter in East Africa, whilst German populations generally migrate through Spain to winter in West Africa. I say generally because in recent years a subpopulation of German blackcaps have started to winter in the UK (UK breeding blackcaps migrate to West Africa). Trapping blackcaps in the UK during winter and then fitting them with leg rings to enable the relocation of the birds on their breeding territories revealed that this population originated in a particular region of central Germany. Berthold compared the migratory orientation of these central German birds with neighbouring southwest German birds in funnel cages, and found that the central German birds oriented westwards towards the UK whilst the southwestern German birds oriented towards Spain. What is more the offspring of both sets of birds, having had no previous migratory experience, orientated in the favored direction of their population. So it would seem to be that in the case of blackcaps migratory orientation is inherited by offspring their parents.

Andreas Helbig has carried out selective breeding experiments to further demonstrate this genetic component of migratory orientation. He took birds from German populations that migrate through Spain and Austrian populations that migrate through Turkey. When he paired birds together according to their

Concept
Zugenrhue

During the period of their migration birds exhibit zugenrhue or migratory restlessness. Under confinement they will attempt to fly in the direction of their preferred migratory route if they are kept in funnel-shaped cages, the wide tops of which provide a view of the night sky (the map the birds use to determine direction).

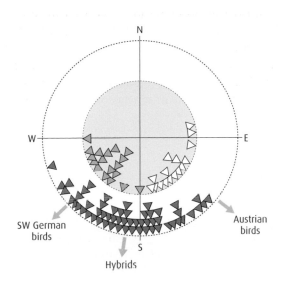

Fig. 4.19 The white triangles show the migratory directions taken by a sample of Austrian blackcaps and the light-coloured triangles show the directions taken by S.W. German birds. Hybrids between the two (dark-coloured triangles) favor intermediate direction. (Average directions are indicated by the arrows outside of the circles.) (From Alcock, J. (2001) *Animal Behaviour, An Evolutionary Approach.* Sinauer Associates, Sunderland, MA; data from Helbig, A. J. (1991) Inheritance of migratory direction in a bird species: a cross-breeding experiment with SE- and SW-migrating black caps (*Sylvia atricapilla*). *Behavioural Ecology and Sociobiology*, **42**, 9–12. Reproduced with permission of Springer-Verlag.)

population of origin he found that their offspring behaved in the way that would be expected and orientated in the same directions as their parents (Fig. 4.19). But look what happened when he created "hybrids" by crossing German and Austrian birds. So the hybrids must inherit a mixture of parental genes that results in some of them migrating in the German manner, and some in the Austrian manner, but on average behaving in an intermediate fashion.

Summary

This chapter has explored the role of genetics in the expression of behavior and discussed the idea that whilst genes do have a role to play, their expression is dependent upon their environment and the environment of the animal. Without an environment there would be no behavior and without genes there would be no behavior. We have seen that although some behaviors are innate and can be performed perfectly first time, behavior is modified as a result of experience. In doing so we have explored a range of types of learning. Finally we have considered the variety of ways in which animals are able to orientate and navigate within their environment, and find distant locations accurately during the course of migration.

Further reading

In her excellent book *Unravelling Animal Behavior* (1995, Longman Scientific and Technical, Harlow, Essex, UK), Marion Stamp Dawkins provides a very readable discussion of the role of genes in behavior. If you want to read more about learning and expand your knowledge base to consider intelligence and cognition, I would recommend *Animal Learning and Cognition: An Introduction* by John Pearce (1997, Psychology Press, Hove). In *Migration: The Biology of Life on the Move* (1996, Oxford University Press, New York), Hugh Dingle provides a very comprehensive review of migration behavior, and in the contributions of various authors in *Spatial Representations in Animals* (edited by Sue Healy, 1998, Oxford University Press, Oxford) you will find detailed discussions of some of the current thoughts on animal navigation and spatial learning.

Communication

Nothing would work in the absence of communication.

Marc. D. Hauser, *The Evolution of Communication,* **1996**

The quotation that I have cited above might at first seem to be a rather grand one – surely there are some things that can work without communication? But I have thought long and hard about these words, as I'm sure Marc Hauser must have done before he committed them to paper, and my conclusion is that if we take a broad definition of communication as being "the sharing of something between A and B" some form of communication is essential in the smooth functioning of all things.

Contents

Key points

◆ In the context of animal behavior communication is the sharing of information between individuals.

◆ Exactly how communication is transferred will depend upon the machinery available to the individual, upon the environment through which the transfer will take place, and upon the evolutionary history of the animals involved.

◆ Although there will be occasions when communicating individuals share information in good faith, there will also be occasions when they attempt to deceive one another.

What is communication?

Dictionary definitions of communication tend to include a phrase along the lines of "the sharing of anything between A and B." Clearly this definition is too broad to be useful to us in the context of a chapter concerning animal communication, but it does provide the basis of a definition that we can use.

So what is shared when animals communicate? Information – when they communicate with one another animals transmit and receive signals, the units of information. But is all information transfer communication? Well the answer to this question will of course depend upon the definition that we choose to adopt. At one level all information transfer could indeed be considered as communication, but in the context of this chapter I want to narrow the definition to include only those examples of behavior that involve the intentional transfer of information between individuals.

A question of intent

In animal behavior terms the use of the word intentional is somewhat problematical. It was never my intention to consider the difficult but fascinating area of animal awareness and consciousness in this book, and so I do not propose to explore the possibility that animals might have intentions that equate to those that we as humans are aware that we have. But I do think it would be useful for us to briefly consider some of the various uses of the words intent. All I ask is that you bear the following distinctions in mind when reading the rest of this chapter.

We could use the word intent in the sense of a voluntary action (voluntary intent), but we might also discuss evolutionary intent – the execution of a behavior that the individual does not choose to exhibit, but which is performed as a result of natural selection. The latter might therefore be described as involuntary intent. To clarify this distinction, consider following examples of human behavior.

Imagine yourself walking down a corridor. You see a friend walking towards you and decide to acknowledge them with a cheery wave of the hand. This is an example of voluntary intent – you chose to perform that particular greeting display. A little further down the corridor you see another acquaintance, not a close friend and not someone who you want to wave at. As you walk

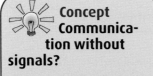

Concept Communication without signals?

It should be borne in mind of course that animals do gain information through the observation of other behaviors that are performed by animals but which have not evolved for this purpose, i.e. they utilize cues.

For example, a pigeon that sees a flock of other pigeons feeding in a field might gain information about the suitability of the field as a feeding site and join them. It did not however gain this information as a result of communication between itself and the members of the flock.

past them however you do make eye contact and your eyebrows involuntarily flick upwards – try it and see if I'm right. This eyebrow flick is an example of involuntary intent because for whatever reason we are adapted to perform this particular signal behavior in this specific situation.

We can also use the word intent in the sense of "what will happen next," and in this chapter we will consider a range of behaviors that are assumed by some to communicate this kind of intent. But once again intent in this case can be used in the voluntary and involuntary sense.

Now I want you to consider your behavior whilst driving a car. If you drive properly you consider the needs of other drivers and will communicate with them. So, for example, if you intend to make a left turn you will let the driver behind know this by applying your left-turn indicator signal. The flashing amber light signals your intent to turn. This is a voluntary signal because you must decide to perform it. However you also communicate unintentionally when you drive. Every time you depress your foot brake red signal lights on the rear of your vehicle are activated. This signal communicates to the driver behind that you intend to slow or stop. But you didn't choose to transmit this information did you? This is an example of an involuntary communication of intent.

Honesty and dishonesty

The red brake light is an honest signal that the car is slowing down and one that should therefore always be taken at face value and acted upon. But the yellow flashing indicator light depends not on the behavior of the car but on the quality of the driver. A good driver will only apply the signal when they are about to turn, i.e. use it honestly, but a bad driver might use it dishonestly or not at all. In this case the signal receiver would do well to make use of other cues when making a decision about whether to act upon the signal or not. We will consider the question of dishonest signals later in this chapter.

Ants and snakes

You may recall that in Chapter 4 we considered the trail-laying/following behavior of some ant species. To recap briefly an ant that has found a food source will lay a very direct pheromone trail between it and its nest that can be easily followed by other

Link
Do the benefits of trail laying outweigh the disadvantages?
Chapter 4

members of the colony. Through this behavior ants communicate information about the location of food to one another. However, the members of at least one other species use the information contained within these trails. *Leptotyphlops* is a snake that uses these trails to lead it to its main prey – ants. Clearly ant-to-ant communication takes place through these trails and through trail laying behavior, but does communication take place between the ant and the snake? In common with a great many authors I would argue that it does not. The trail has been produced with the primary function of facilitating information transfer between ants, and the behaviors of all of the ants involved are adapted to maximize the efficiency of information sending and receiving between them. The trail was not however laid down to communicate the presence of the ants to the snake. The snake does make use of information encoded within the trail, but that information was not intentionally transmitted to the snake. And this is the important point. In the context of this chapter I would therefore define communication as, "the voluntary or involuntary sending and receiving of signals, that are likely to have evolved for this purpose, between individuals that are committed to maximizing the efficiency of the transfer of information between them." Take note that this definition does not make assumptions about the honesty or otherwise of the information that is transferred. Also note that in the definition I refer specifically to signals as behaviors that are likely to have evolved to enable communication.

The evolution and design of signals

Animals communicate with one another in a bewildering variety of ways. Some animals use sound and others use chemicals. Some animals communicate by touch or vibration and some use visual signals and body postures. In this section of the chapter I want to consider a number of examples of communication and the evolutionary and environmental factors that have shaped the signals that they involve.

Frog semaphore

Precisely which methods of communication are used by any one animal species will depend in a large part upon the environment in which they communicate (see below). But the method used

Concept
Selective pressures and the evolution of signals

It seems clear that a number of selective pressures operate during the evolution of a signal.
• Not all signal types will stimulate the sensory systems of the receiver to the same extent.
• The environment through which signaling is to occur will strongly influence the design of a signal – after all not all signal types will travel from sender to receiver with equal ease.
• The interests of the signaler and receiver will not always coincide, selective pressures might favor a dishonest signal on the part of a sender, but similar pressures acting on the standpoint of the receiver might favor honesty.

must also depend to an extent upon the equipment available to the individual.

Most species of frog and toad croak at one another as a form of social communication. However one species, the Panamanian golden frog *Atelopus zeteki* lacks an essential component of the machinery required for acoustic communication. This frog has no tympanic middle ear and so it is to a large extent deaf. Golden frogs do respond to sound, so they are not completely deaf, but the available experimental evidence strongly suggests that compared to their fully hearing cousins they do have reduced acoustic sensitivity. Add to this the fact that they inhabit fast-flowing mountain streams where high levels of ambient noise would make it difficult for any animal to hear a signal, and the extent of the unsuitability of an acoustic communication system for this animal becomes apparent.

So do golden frogs live a life of half-heard and confused messages? No they do not. During their evolutionary development an alternative mode of communication has arisen – one that makes use of machinery that is already available them.

We have already seen in Chapter 2 that amphibians have a highly developed visual system, and one that is particularly sensitive to movement. It should not be a surprise therefore that these animals have developed a visual communication system. Researchers working in the Panamanian cloud forests have established that golden frogs communicate with one another using a system of semaphore. Individuals wave at one another, and although the exact messages that they transmit remain elusive to us, they are thought to relate to territorial defence and sexual selection – the same functions that have been attributed to the croaks of their cousins.

Sensory preferences and signal design

Some male insects are tricked into pollinating the flowers of certain orchid species because those flowers bear more than a passing resemblance to female insects. This does not happen because the plant is in itself attractive to the insect. The

---

Focus on signal evolution, ritualization, and antithesis

The evolution of a signal can be thought of as a process having six basic steps:
1. An association between a behavioral, physiological or morphological cue and a behavioral motivation/condition as an incipient signal on the part of the sender.
2. The perception of the cue on the part of the receiver.
3. The relating of the cue to the motivation/condition of the sender by the receiver.

4. The development of a decision rule by the receiver in response to it relating the cue to the motivation/condition.

5. The evolution of a response to the cue/signal by the receiver.

6. The process of ritualization refines the cue to create a true signal. As an example take the blue tit visual threat display illustrated in Fig. 5.4. How does a display involving an exaggerated crouch with the tail fanned and lowered and the wings spread communicate "I am likely to attack" to the signal recipient? If we follow the six steps outlined above we can easily imagine a scenario in which this could happen.

Imagine that prior to the evolution of the signal birds attacked one another without preamble. Blue tit attacks, when they do occur, involve a lunge or leap towards the opponent, the initial moment of which involves the attacker crouching. So a crouch in an agonistic context is a cue that an attack is about to be launched (step 1). If the recipient recognizes the crouch and links it to the fact that it precedes an attack (steps 2 and 3), it could use this relationship to develop a simple rule – "if I see you crouch you will attack me" (step 4) and evolve a coping strategy – flee (step 5). Through ritualization the crouch is exaggerated until it becomes the display that we recognize (step 6).

Intention movements are only one example of a cue with the potential to become a signal. The threat display of many fish species involves exaggerated gill raising; it seems likely that this display has evolved from a morphological/physiological cue. During a fight an animal's metabolic rate increases and so does its rate of oxygen intake. In the case of a fish this is apparent as more rapid or wider gill opening.

Ritualization does not necessarily have to involve the elaboration of the cue through its exaggeration. It could involve its simplification, or an increase in its performance as a repetition. But whichever form it takes the outcome of ritualization will be a reduction in the ambiguity of the signal, a clear benefit to both of the communicating parties.

When two signals communicate very different messages there is a tendency for them to evolve to be as different from one another as possible. This phenomenon is termed **antithesis**. So for example whilst the blue tit threat display involves a flattening of the crest feathers, the opposing submission display involves holding them erect.

Traditionally when we have examined the evolution of signals we have tended to focus upon the signaler. But as we will see later in this chapter it is important to remember that the receiver has just as big a part to play. Recently it has been recognized that exactly which behaviors act as cues and become signals may depend upon the sensory biases and preferences of the signal receiver.

plant is in fact exploiting a pre-existing bias in the sensory system of the insect – the fact that male insects find female insects attractive. In terms of signal design this process is termed **sensory exploitation**, the theory that signals evolve to exploit pre-existing sensory biases on the part of the receiver. In a similar way the courtship display of the water mite *Neumania papillator* is thought to have evolved to exploit the fact that female mites are especially sensitive to the water-borne vibrations emanating from their copepod prey. During the display male mites wave their trembling forelegs as they approach females, replicating the vibrations of the copepod.

The environment and signal design

Channels of communication

The pathway linking a signaler and the receiver of the signal is known as a channel. The term **channel of communication** is used to describe the various sensory modalities that are utilized by animals to facilitate effective communication. Some such as the **electrical** sense are relatively rarely used because they can only operate in one specific environment (in this case water). Others are used more widely, and there are four in particular that are commonly used in a range of environments and by a wide range of animal taxa. These are the **visual**, **auditory**, **chemical** and

Male swordtail fishes of the species *Xiphophorus helleri* possess a long extension on the base of their caudal fin. This is known to be a sexual signal and when given the choice, females of the species show a marked preference for well-endowed males. There are a number of species in the genus *Xiphophorus*, some sporting swords and some of which are swordless (Fig. 5.1). Swordlessness is known to be the ancestral state in these fish, and so within the genus it is possible for us to identify pairs of species that diverged from one another prior to the evolution of the sword. This situation has provided researchers with an almost unique opportunity to examine the evolution of a signal. Two of the questions that they have focused upon are what is it about having a big sword that makes a male attractive, and which came first – the sword or the female preference for it? At first glance this might seem to be a rather strange question to ask. After all, how could they prefer something that did not yet exist?

Which came first?

Gil Rosenthal and Christopher Evans have shown that female *X. helleri* prefer males with swords when they are presented with a choice (Fig. 5.2). But it might surprise you to find out that if instead of using *X. helleri* in their experiments they had used the swordless species *Xiphophorus maculatus*, and they had given these females the opportunity to choose between a normal swordless male and a male with a fake sword, the result would be the same. The females would find the sword attractive.

As the swordtail evolved in the genus *Xiphophorus* more recently than the evolutionary divergence of the ancestors of *X. macualtus* and *X. helleri*, it would seem that the sword evolved to exploit a pre-existing sensory preference on the part of the female.

Why is the sword attractive?

What is it about the signal that makes these males attractive? Through a series of ingenious experiments during which female fish were offered the opportunity to associate with images of fish and parts of fish presented to them on a video monitor, Rosenthal and Evans have established that the sword in itself is not what makes a male attractive. In fact it would appear that the signal serves to increase the apparent size of the male, and that it is size that the females are attracted to (Fig. 5.3).

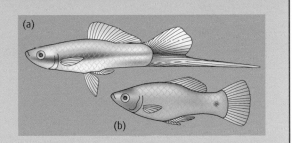

Fig. 5.1 Male swordtail fish of the species *Xiphophorus helleri* have a long sword extending from the base of their caudal fin (a); male platyfish *Xiphophorus maculatus* do not (b).

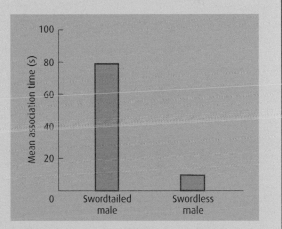

Fig. 5.2 Females of the swordless species *X. helleri* prefer males with swords when given a choice. (From Rosenthal & Evans 1998.)

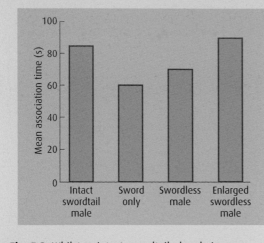

Fig. 5.3 Whilst an intact swordtailed male is more attractive than a standard swordless fish or a detached sword, females appear to have the greatest preference for an artificially enlarged swordless male. (Data from Rosenthal & Evans 1998.)

Female preference for large size may result incidentally from greater stimulation of the sensory system, or it may have an adaptive significance. Bigger males might make better mates in some way. The evolution of the tail has allowed males to exploit the sensory bias of the females without necessarily increasing their overall body size. It has been shown that male *X. helleri* will invest in both tail growth and body growth when they have a plentiful food supply. If food is restricted, however, investment in body length ceases, but investment in sword length does not. So the sword might offer males a metabolically inexpensive way of appearing larger than they are.

Rosenthal, G.G. & Evans, C.S. (1998) Female preference for swords in *Xiphophorus helleri* reflects a bias for large apparent size. *Proceedings of the National Academy of Science USA*, **95**, 4431–6.

Plate 5.1 The lionfish warns off would-be predators with a display of its ferocious looking spines. © C. Waller.

mechanical channels. Each of these channels presents advantages and disadvantages to the user, and these may vary from environment to environment. For this reason individuals may rely upon a combination of them.

The visual channel

The visual channel offers distinct advantages in terms of communication in that visual signals are easily locatable, the receiver can literally see where they come from (Plate 5.1). Another clear advantage of visual signals is that they can be changed rapidly to allow rapid changes in the message that the signaler wishes to convey (think for example about human sign language and the amazing volume of information that can be transferred from individual to individual by it). However visual signals cannot travel round corners, or through dense vegetation, their high level of locatability might result in a heightened risk of predation for the signaler, and they are ineffective in darkness, so the visual channel does have some drawbacks. Most visual communication is therefore carried out over short ranges.

Blue tit agonistic communication

Visual communication is very important to blue tits *Parus caeruleus* when they forage in flocks during the winter months. When two blue tits meet at a food source during the winter, they may feed amicably side by side. If however the food item that they have found is not easily divisible, they are likely to fight over it. Fights between blue tits are common, but you might easily overlook them if your idea of a fight is something along the lines of a human experience of physical contact through kicks and punches. Physical fights do occur between blue tits, but they are very rare. In fact during a program of observations that involved close to 3000 agonistic contests (as fights are known in the behavioral jargon) between birds, I saw less than 30 of them. I expect that based on this observation the same two questions that immediately occurred to me will have occurred to you – why are physical fights rare? and, how were the other 2970 or so contests resolved?

The answer to the first of these questions is an intuitively obvious one. As physical fighting is potentially costly in terms of energy expenditure and risk of injury we would expect natural selection to favor the evolution of contest resolution mechanisms that avoid it. The answer to the second question is in two parts. Some blue tits will avoid contests completely by simply giving way to one another. For example, a female bird will almost always abandon a food source when a male arrives. In this context it would appear that males have dominance with respect to females. And in the jargon of the field we might say that the male supplanted the female. But when two birds of the same sex meet this simple rule cannot be used. In such a situation blue tits achieve contest resolution through the performance of ritualized visual displays (Fig. 5.4). Interacting birds posture to one another during contests and it is generally assumed that the birds use information encoded in these displays to make decisions about when to quit the field and when to continue the contest (Plate 5.2).

As an aside we might expect the selective value of resolving disputes without physical fighting to increase as the potential the combatants have to do real harm to one another increases.

Key reference
Scott, G.W. & Deag, J.M. (1998) Blue tit (*Parus caeruleus*) agonistic displays: a reappraisal. *Behaviour*, **135**, 665–91).

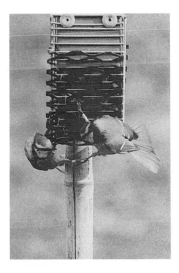

Plate 5.2 Blue tits engaged in agonistic display. Both birds have flattened crests and erect nape feathers; one bird also has its tail feathers fanned. © G. Scott.

Fig. 5.4 During agonistic encounters blue tits (*Parus caeruleus*) adopt a range of postures. Submissive birds raise the feathers of their crest and puff out their bodies to appear round (a). Threatening birds hold themselves erect (b), pointing to the sky, or crouch (c,d) with their crest flattened and nape feathers erect and their wings held away from the body (b,c) or even flapped above/behind it (d). They may point their bills at the ground (c) or open them towards their opponent (d). Often threatening birds fan their tails (b–d).

This is exactly what James Serpell found when he carried out a review of the agonistic behavior of the members of the lorikeet genus *Trichoglossus*. Large lorikeet species with heavy and powerful beaks (the main weapon used in a physical fight) use up to 20 highly ritualized display behaviors during contests and they are unlikely to resort to a physical attack. Smaller species, however, with small beaks and therefore relatively little capacity to do real harm to one another have a far less well developed display system and exhibit a far higher tendency towards physical fighting.

The acoustic channel

Many marine mammals have highly developed visual systems and so we might expect them to communicate visually. But light in the sea attenuates (is absorbed and/or scattered) very quickly, and so even in the clearest waters marine mammals are unlikely to be able to see an object clearly if it is more than a few meters away. Sound, on the other hand, attenuates more quickly in air than it does in water, and the acoustic signals of whales are thought to have effective ranges of more than 100 kilometers. So it is far from surprising that those mammals that have returned to a marine existence have evolved highly specialized means of acoustic communication. Acoustic signals in general are a very effective means of communication in both the terrestrial and aquatic environment because they offer rapid signal transmission and a high potential for rapid signal change.

As we have discussed in the case of the Panamanian golden frog, however, they are subject to environmental interference, and changing levels of background noise, or sound-dampening vegetation structure, might have a negative impact upon their effectiveness. There is evidence that some species are able to take account of these environmentally imposed problems and modify their sound signals accordingly. Humpback whales are known to alter the frequency at which they sing during different times of the year. It had been assumed that this change in song equated to a change in the priorities of the singing individuals. Perhaps during one part of the year the songs were related specifically to courtship, whilst in another season they were related to the availability of food. Recently, however, an alternative to this hypothesis has been proposed. By modeling the dynamics of sound transmission researchers have found that different frequencies are propagated through water most effectively at different temperatures, specifically lower frequencies travel best through warmer water and higher frequencies travel best through colder water. So if whales are behaving in a way that would maximize the effectiveness (range) of their songs, we should expect them to use lower frequencies in summer than in winter, and from the available observations that is exactly what they seem to do.

With the exception of human speech the auditory communication system that is most familiar to us must be that employed by the song birds. Males of these species (although not exclusively as there are some accomplished female singers) devote considerable time and energy to the development and production of their songs, and a considerable effort has been made on the part of a large number of behaviorists to understand their significance (Plate 5.3).

One key hypothesis to explain this behavior suggests that male song has a mate attraction function. Evidence to support this idea comes from various observations and experiments including the following. Firstly, singing peaks during the period at the onset of the breeding season when males and females are pairing up. Secondly, males of monogamous species reduce their singing activity once a mate has been acquired, but by contrast the males of polygamous species carry on singing after they have acquired a partner. The latter of course pair with more than one female. This effect can be further explored experimentally if the mate of a

Plate 5.3 Is the song of this robin aimed at potential mates or potential rivals? © G. Scott.

Pygmy marmosets (*Cebuella pygmaea*) are small primates that are restricted to river-edge forests in tropical South America. Visibility in these forests is poor and so pygmy marmosets, like so many other forest species, depend to a high degree upon vocal signals when communicating with one another. Communication is particularly important to these animals, living as they do in small, stable groups, typically of less than 10 individuals.

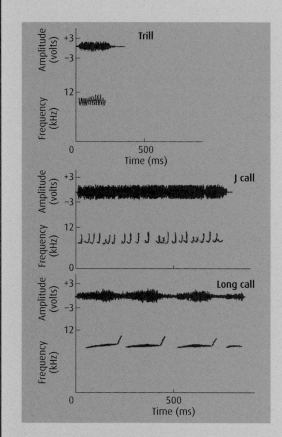

Fig. 5.5 Sonograms of the Trill, J call, and Long call of pygmy marmosets. (From de la Torre & Snowdon 2000. Reproduced with permission of Elsevier.)

The members of the group cooperate in the defence of their small territory and in the raising of young. In such a context vocal signals that facilitate group cohesion are particularly important.

Included within the vocal repertoire of the pygmy marmoset are a class of vocalizations termed contact-calls, the function of which is to maintain short-range contact between individuals and thereby facilitate group cohesion and interaction. But pygmy marmosets routinely use three different contact calls, the "Trill", the "J call", and the "Long call." These calls all consist of rapid pulses of relatively high frequency sound (in the 7–12 KHz range), but as can be seen from Fig. 5.5, their acoustic characteristics do vary over time.

Stella De La Torre and Charles Snowdon have studied the contact calls of two wild populations of pygmy marmosets in the riparian forests of Amazonian Ecuador. Through their experiments and observations they have shown that all three of the calls are used when animals are foraging, traveling, or resting. So in a behavioral context the calls appear to be equivalent. Why then are there three of them? There is one context in which an individual is much more likely to use one call rather than another (Fig. 5.6). Specifically the Trill seems to be most often used when the distance between the signaler and receiver is small, no more than 10 meters or so. The Long call on the other hand is most commonly used when the signaler to receiver distance is greater, and the J call is most commonly used at an intermediate distance. So these three calls can be used by individuals to communicate to one another information about the level of dispersal of the group.

But why these three particular calls? Is it an accident that the Trill signals low dispersal – why doesn't the Long call do that job? To answer these questions the researchers played recordings of the calls in typical marmoset habitat and then measured various parameters of them at increasing distances from their source. In this way they were able to examine the different transmission levels of the three calls through the environment. As expected all three calls suffered some level of degradation as they traveled through the forest, but they did

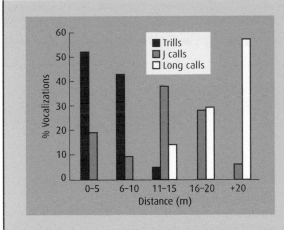

not all suffer to the same extent. The Long calls were the least affected and they were the only calls that were still audible at 80 meters from the playback source. Trills and J calls could both be detected up to a distance of some 40 meters, but the former had suffered significantly more degradation at this distance and in effect their integrity was only maintained to a distance of about 20 meters. So it seems that selective pressures relating to the physics of sound transmission and habitat acoustics have resulted in the evolution of three distinct contact calls in this species.

De la Torre, S. & Snowdon, C.T. (2002) Environmental correlates of vocal communication of wild pygmy marmosets, *Cebuella pygmaea*. *Animal Behavior*, **63**, 847–56.

Fig. 5.6 Trills are used by animals close to one another and Long calls are used by distant animals. J calls are used at all distances, but predominantly in cases of mid-range isolation. (From de la Torre & Snowdon 2000. Reproduced with permission of Elsevier.)

monogamous male is removed from his territory shortly after they have paired up. If singing attracts mates we would expect this male to burst into vigorous song, and this is exactly what happens. His singing will only subside if his mate is returned to him, or he finds a replacement for himself.

The fact that the song of the male attracts the female is only one part of this story. In Chapter 8 we will see that it pays a female to be choosy when it comes to accepting a mate. Songs are thought to provide the female with a means by which males can be compared. Generally females prefer males that sing more complicated songs and those with the stamina for lengthy singing bouts. These males tend to be healthier individuals, with better territories and enhanced parenting potential. They may also offer the female high quality genetic material that they will pass on to their offspring. We will consider these possibilities in more detail in Chapter 8.

Experimental evidence does suggest, however, that mate attraction is not the only function of bird song. If I were to capture and remove the males from two territories, but then to "replace" one

Link
When choosing a mate females should compare males and males should compete for their attention.
Chapter 8

of them with a tape-recording of his own song, nonterritorial intruders would quickly fill the gaps that I had created. But in almost every case it would be the territory without the tape that was filled first. So something about the song of another male deters intruders. It is clear that not all members of a population will react to the same signal in the same way, an important consideration when studying communication.

The electrical channel

In murky shallow water neither light nor sound are particularly effective as a medium for communication. In these conditions two groups of teleost fish, the African mormyriformes and the South American gymnotiforms, have independently evolved a system of communication which involves the electrical channel.

Electrical signals cannot be effectively transmitted through air and so the use of this channel is restricted to the aquatic environment. Even here it is only useful at a range of up to a meter.

Within these limitations, however, there is observational and experimental evidence that these fish species are able to use the pattern of electrical discharges and interdischarge intervals, together with the characteristics of the discharge itself, to communicate information about species identity, the sex of an individual, and dominance relationships between individuals. Male *Sternopygus macrurus*, a gymnotiform species, add periodic modulations to their usual backgroud pulse pattern in an attempt to attract and impress females. These courtship songs are a clear and interesting parallel to the songs of whales and of birds in that they are complex and costly to produce.

The chemical channel

The advantages of chemical communication are not as immediately apparent as those offered by the visual and acoustic channels. To be effective the molecules that make up a chemical signal need to move from the sender to the receiver. In some cases this is achieved by the sender physically placing the signal onto the receiver, or by the receiver moving to a deposited signal to pick it up. More often it involves the movement of the chemical through air or water by the process of diffusion. Natural or signaler-produced currents may help to move the signal in

Plate 5.4 After the golf ball was coated with a female sex attractant pheromone this male crab proceeded to attempt to mate with it! © J. Hardege.

an appropriate direction, but this reliance on diffusion means that chemical signals typically follow irregular routes and travel at reduced speeds (while a sound pulse can travel from a signaler to a receiver in milliseconds, chemical transmission may take minutes, hours, or even days). Reliance on diffusion also means that chemical signals do not lend themselves to a set temporal pattern and that they are not a suitable medium through which to communicate a rapidly changing message (Plate 5.4).

In the context of the animals that use them, however, the advantages of chemical signals are obvious. Take for example the pheromones deposited by the trail-laying ants that we discussed in Chapter 4. These chemicals provide an unambiguous message to the signal receiver, and one that will persist in the environment for a sufficiently long period for it to be useful to a number of individuals (a sound signal or physical expression would not). For similar reasons they offer territorial animals a means of signaling ownership of an area without the animal needing to be physically present. However, the fact that these chemicals are volatile does mean that the signal will disappear when its usefulness wanes, an advantage to the ant – but not to the defender of a territory. To overcome this problem not all chemical signals have evolved to be similarly ephemeral, some provide a longer lasting signal than others.

The beetles of the genus *Carpophilus* have recently become a major agricultural pest in parts of Australia. They attack ripening stone-fruit (nectarines and peaches for example) and are a vector of brown rot, another threat to the industry. In those areas where beetle infestations become sufficiently large entire crops can be lost and so the economic implications of this pest are clearly significant. One solution to the problem might be to saturate affected orchards with an insecticide, and of course at one time that would have been our knee-jerk reaction. But thankfully in the modern world we are at least less likely to consider this option, and in the case of the ever-growing organic industry it simply is not an option. There is today a more elegant weapon in our armoury. David James and his team at Washington State University have found a way to use the beetles' own communication system against it.

Male *Carpophilus* beetles secrete an aggregation pheromone that is irresistible to members of their own kind. The chemistry of the pheromone is well known, it can be readily synthesized, and the synthetic product is effective as an attractant at distances of up to at least 500 meters. If traps (simple funnels that beetles slide through and into a water-filled container) are baited with a small amount of the synthetic pheromone, the beetles can be easily caught.

Putting traps directly into fruit trees is not however an ideal strategy. If that is done there is the chance that the beetles will still cause some damage before they are caught. So James and his team placed traps on poles sited 5 meters outside of an orchard, and at 3-meter intervals all of the way around its perimeter. A perimeter-based suppression protocol as this arrangement is known has the advantage that in addition to drawing pests out of the orchard it also catches the pests that are on their way in.

But does it work? Well in one of their experiments the researchers trapped a staggering 94,240 beetles in 54 traps during the 4-week period prior to fruit harvesting, and a further 14,200 during the 4-week harvest period. Very few beetles were found inside the orchard, and no damaged fruit were reported. Clearly the technique is a success.

The principal of using a pest's own communication system as a weapon against it is not new, nor is it restricted to the control of fruit pests. The same idea is at the heart of a number of initiatives to control a range of stock pests and to control a range of insects that present a risk to human health, either directly or as a result of the agents of disease that they transport.

James, D.G. *et al.* (2001) Pheromone mediated mass trapping and population diversion as strategies for suppressing *Carpophilus* spp. (Coleoptera: Nitidulidae) in Australian stone fruit orchards. *Agricultural and Forest Entomology*, 3, 41–7.

The mechanical channel

The final main mode of communication that I want to consider is the mechanical channel. This is generally taken to include communication that is facilitated via direct contact between individuals (touch) and through the transmission of information via substrate-borne vibrations. Clearly communication by touch is effective only at extremely short range, and it is likely that the range of substrate-borne vibrations is also small. After all these signals will degrade as they are transmitted and may well be masked or confused by other environmental vibrations. The mechanical channel is extensively used for short-range communication by a wide range of invertebrate species, to which it offers the advantages of high locatability and high rate of signal change. It is important, for example, during the courtship behaviors of web-making spiders. As ardent males approach potentially dangerous females they tap out species-specific signals on the strands of the web to lessen the chances that they will be mistaken for prey.

Alarm calls

One of the most commonly produced calls of the European robin (*Erithacus rubecula*) is a shrill whistle or "seet." This call has a very specific context. Its use is reserved for situations when the robin has detected a potential predator. It is an alarm call. One function of alarm signals (which may involve any of the communicatory channels, not just the auditory one) is likely to be that they communicate to the predator that it has been detected, that the element of surprise has been lost, and that to pursue an attack would be a waste of effort.

Link
Communicating with a predator might deter an attack.
Chapter 7

A second function of alarm calls is that they enable individuals to share information about predator proximity and the risk of attack. In the case of the "seet" call this sharing is very wide indeed. All of the songbird species commonly found in the robin's woodland habitat have a similar call and each will respond to the calls of the others. In fact the "seet" calls of these species are so very similar that on the occasions that I disturb mixed-species flocks and release a cacophony of alarms I can not tell the "seet" of a robin from that of a chaffinch or a blackbird. It is thought that one of the main selective forces shaping the acoustic qualities of these calls is their level of locatability by the predator. After all in many circumstances it would not pay to reveal one's whereabouts to a hungry hunter. But because "seet" calls have a high frequency and a narrow frequency range, and because they are a continuous noise without interruptions (Fig. 5.7), it is thought that it is very difficult for the predator to get a bi-aural "fix" on them. If the predator cannot lock-on to its target it cannot attack it. But as we will see in Chapter 7, sharing responsibility for alarm giving and triggering a cacophony of alarms and a scramble of escaping individuals will both benefit the hunted animal greatly.

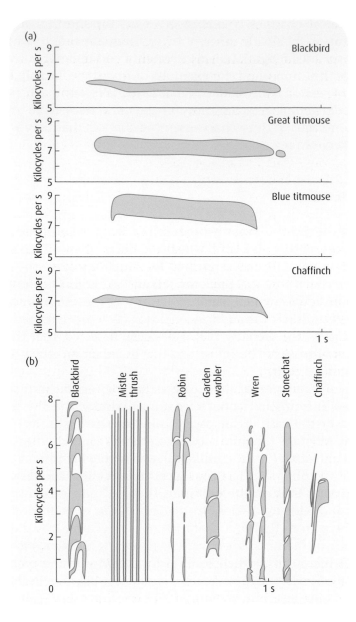

Fig. 5.7 Sonograms of the alarm calls of four species of bird commonly found in the same habitat (a), and mobbing calls of seven species that overlap in both habitat and mobbing target (b). Notice that the frequency range of the alarm calls is narrow, making them difficult to localize. In contrust the sources of mobbing calls are easily detected by the target – they need to be. (From Marler, P. (1959) Developments in the study of animal communication. In *Darwin's Biological Work*, ed. by Bell, P.R., pp. 150–202. Cambridge University Press, Cambridge. Reproduced with permission of Cambridge University Press.)

Crying wolf

In the forests of the American tropics mixed-species flocks of foraging birds are common too. As we would expect the members of the flock "trust" one another and make use of a shared alarm call.

However, there are occasions when it would appear that this trust has been misplaced. In any flock there is usually one species that occupies a lead position. This is often a shrike-tanager feeding high in the canopy and encouraging the rest of the flock to follow it by means of a distinctive contact call. While the other flock members are foraging diligently in the understory below it, the shrike-tanager tends to adopt a perch-and-search strategy. It waits until it spots a prey item disturbed by its flock mates and then pounces on it from its vantage point.

From these elevated vantage points these birds are perfectly placed to act as a look out for the whole flock because the main threat to these birds comes from above in the form of hunting hawks. When the shrike-tanager sees a hawk it will invariably sound the alarm and all of the flock members will respond appropriately. They will stop what they are doing (but hold on to any prey item that they have just caught) and either look up to ascertain the likely direction of attack, or just break for cover. But occasionally the shrike-tanager will cry wolf. If it sees a particularly choice prey item that has been flushed by the flock, but one that another bird is likely to get to before it can pounce, it will sound a false alarm. Then while the rest of the flock are engaged in evasive actions the deceiver will swoop down and claim its prize.

There is an added refinement to this behavior. If the shrike-tanager initiates a false-alarm call too late and the desired prey item is snapped up by one of the other birds in the flock, the call is aborted. The shrike-tanager will change it into another vocalization mid-call. Presumably this flexibility helps the bird to retain the trust of its flock mates.

Levels of risk and referential signals

The members of some species vary their alarm call depending upon the level or nature of their perceived risk. The robin "seet" call, for example, is a higher-risk warning than is the mobbing call of the same species. Most ground squirrel species use different alarm calls in different risk contexts too. For example, the threat of attack from the air is usually communicated via a short, shrill whistle, whereas the presence of a terrestrial predator is more commonly indicated by a trill or a chatter. These two types of attack do have different levels of risk associated with them. Attacks from the air are usually fast and likely to be detected at the last minute – they present a high level of risk. A predator on the ground, on the other hand, is usually slower moving and detected at a greater distance, therefore posing comparatively less risk. One question that this kind of observation presents is what is the nature of the information encoded in the alarm? For example in its shrill whistle is the ground squirrel communicating the message, "High-risk predator!" or is the message "Hawk!" In this case at least it seems that the former is the most likely because a distant hawk will elicit a trill and the sudden appearance of a coyote or weasel in the ground squirrel colony will result in a chorus of whistles.

There are however some alarm calls that do seem to convey very specific information about the nature of the predator that has been detected. When a young vervet monkey spots a bird in the sky above it, it will give an alarm call. In this case a sort of "cough-cough" noise. At this stage the call appears to be an innate possible-danger-above signal because it is given as a response to any large flying object dangerous or otherwise. But as the monkey matures the range of stimuli that will trigger the call narrows, presumably as a result of the individual learning what might be a threat and what might not. Eventually the use of this alarm call will be restricted to those situations when an eagle is spotted in the skies above. Upon hearing the call the members of the troop will scan the sky to locate the threat and then make a dash for the cover provided by dense vegetation.

Eagles are only one of the predators that attack vervet monkeys. They commonly fall prey to leopards and to snakes. The appropriate escape response when threatened by each of these predators is different, as are the alarm calls that release them. On detecting a leopard the call that is given is a "bark" and the escape response is to climb the nearest tree and move to branches that are too thin for the leopard to follow. But snakes elicit a "chutter" call and mobbing behavior by the troop members to drive the threat away.

Experimental evidence that the alarm calls of these and other species of monkey really do provide information about the precise nature of the predatory threat, and can therefore be considered as being referential signals, has been provided by a number of researchers including Cheney and Seyfarth. They took a tape recording of the eagle alarm call of a particular vervet monkey and played it repeatedly to a troop of wild vervets. As one would expect the monkeys initially took heed of the call and scanned the sky for an eagle. But of course they didn't find one. As this

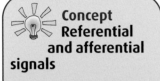

**Concept
Referential
and afferential
signals**

Many of the signals that are transmitted by animals are afferential – they communicate some information about the sender itself.

Referential signals, by contrast, communicate information about an entity that is external to the communicating individual. They enable animals to communicate with one another about the things around them.

process was repeated the troop eventually habituated to the obviously unreliable call and ignored it. Once habituation had occurred the researchers did one of two things. To one troop of habituated animals they played the same eagle alarm but one recorded from a different individual. To another group they used a recording from that animal to which they had habituated, but this time they played the leopard alarm. In both cases the troop members dishabituated. When the caller changed they looked to the sky again. Presumably they were able to recognize that the identity of the caller had changed, and because they had no information about the reliability or otherwise of this new signaler they were forced to take the message seriously, and look for an eagle.

When the call changed but the unreliable caller was retained the members of the troop dashed for the tree-tops. Evidently they had recognized that the nature of the threat had changed and having no evidence that an animal proven to be an unreliable eagle-spotter would also have poor leopard recognition skills, they responded appropriately.

Syntax?

Referential alarm calls are also used by the Diana monkey (*Cercopithecus diana*) and by Campbell's monkey (*Cercopithecus campbelli*). These two species often live along side one another and both are attacked by the same predators. Each species has a referential eagle-alarm call and a referential leopard-alarm call. Acoustically the calls of the two species are very different, but as we might predict each takes note of the calls of the other. So when a Diana monkey hears the leopard-alarm of a Campbell's it will respond with its own leopard-alarm call and then head for the trees. During his investigations of the vocal repertoires of these two monkey species Klaus Zuberbühler has arrived at a startling conclusion. It is highly likely he feels that these monkeys may exhibit lexical syntax.

Syntax describes the rules governing the order in which signals are given. You and I follow syntactic rules when we communicate. If we did not I doubt that we would understand one another very well if at all. Two classes of syntax are recognized, phonological and lexical. The former refers to a simple rule of order – the way that signals are built onto one another to create a bigger signal. Phonological syntax is commonly observed in studies of

Link
Animals can learn to ignore one another.
Chapter 4

Key reference
Zuberbühler, K. (2002) A syntactic rule in forest monkey communication. *Animal Behaviour*, **63**, 293–9.

animal signal systems. But lexical syntax is more interesting, perhaps because it is such a striking characteristic of our own language system and because direct evidence for it in nonhuman animals is rare. Lexical syntax refers to the rules whereby the use of signals in conjunction with one another results in a new signal that conveys a new message and one that is independent of the meanings of the component signals. Zuberbühler noted that under some circumstances Campbell's monkey-alarm calls are preceded by a distinctive "boom-boom" call. Contextually these calls are heard when something startles the monkey that could indicate the presence of a predator – a large branch breaking for example. He also noted that whereas Diana monkeys will always react to a Campbell's-alarm as an alarm, they do not respond to a boom-boom-alarm sequence in the same way. In fact they ignore it.

So what is happening here? Well it could be that the boom-boom call simply acts as an inhibitor, preventing the expression of the typical behavioral response to an alarm call stimulus. To test this hypothesis Zuberbühler exposed wild Diana monkeys to one of two tape recordings (Fig. 5.8). When he played the boom-boom-alarm sequence of the Campbell's monkey the Diana monkeys failed to respond (as we would predict given the field observations that had already been made). But when the sequence changed and the Diana monkeys heard the Campbell's boom-boom followed by their own alarm call they responded very differently. This time the Diana monkeys apparently ignored the boom-boom, took notice of the alarm call, and responded by alarming and fleeing. Based on these results it seems unlikely that the boom-boom is a simply inhibitor.

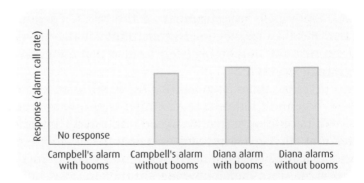

Fig. 5.8 Diana monkeys always respond to their own alarm call even if it is preceded by a Campbell's monkey boom-boom call, but they will only respond to the Campbell's monkey alarm when it is not preceded by the boom-boom. (From Zuberbühler, K. (2002) A syntactic rule in forest monkey communication. *Animal Behaviour*, **63**, 293–9. Reproduced with permission of Elsevier.)

So it seems that to a Diana monkey the boom-boom call is only relevant if it precedes a Campbell's alarm and that in this context it acts as a modifier changing the message conveyed by the signal from say "Leopard!" to something else, perhaps something like, "Possible leopard – but no immediate danger". At the moment however we can probably only speculate on the complexity of the communicatory abilities of our primate cousins. But as our knowledge base increases I am sure that they will continue to amaze us.

Individual recognition; what's in a name?

One conclusion that may be drawn from the experiments carried out by Cheney and Seyfarth is that Vervet monkeys are able to discriminate between the alarm call of one individual and that of another. They can tell one another apart. This ability has been documented in a very wide variety of species. Bull frogs are able to discriminate between the croaks of their neighbors and of strangers, territorial birds compare one another's songs in the same way. Also turnstones have been shown to be able to recognize individuals on the basis of their facial plumage patterns. Mammals (including humans) can identify individual members of their own species by smell, on the basis of differences in the breakdown products of the proteins coded by the genes of the MHC (major histocompatability complex) and which are secreted in sweat and present in urine.

The ability to recognize one another is of vital importance to social animals, animals that are territorial, and animals that interact with one another on a regular basis. Fights might be avoided if you recognize an individual by whom you have been beaten recently. There may be no need to mount a defensive action against an animal singing close to your territorial boundary if it is just the territory holder next door, but if it is a stranger the situation may be very different.

This kind of recognition can be achieved if individuals compare one another's physical properties, or the details of one another's signals. We do this when we recognize each other by voice or facial features. But we also recognize one another in another way. We use the abstract labels that we refer to as names. We might think of our species as being unique in its use of names, but there

Plate 5.5 These bottle nose dolphins rely upon their signature whistles to stay together. © G. Scott.

is at least the possibility that we share this ability with another group of species – the dolphins.

Bottlenose dolphins can be taught to use a vocalization to label an object and then to use that specific sound to report on the presence or absence of that object during experimental trials. It seems likely therefore that they would have the capacity to recognize and label one another in a similar way. Furthermore, it seems safe to assume that individual recognition must be essential for these animals given their lifestyle and environment. Young dolphins remain with their mothers for up to 8 years, and although dolphin "societies" are relatively fluid with animals coming and going on a regular basis, there exist within this apparent chaos very stable subgroups of individuals that remain together for prolonged periods. For these animals maintaining contact in the sea must involve acoustic communication because the individuals involved are unlikely to be able to see one another through more than a few meters of seawater (Plate 5.5).

One striking feature of their varied vocal repertoire is the development by all dolphins during the first few months of life of a highly distinctive and extremely stable vocalization that is referred to as a **signature-whistle**. Each animal has its own signature-whistle and it will remain relatively unchanged throughout the animal's relatively long lifetime. Research has also revealed that dolphins can remember at least eight different signature-whistles and playback experiments suggest that they can use them to reliably discriminate between individuals. The whistles are used in a variety of behavioral contexts, but during studies of captive animals Vincent Janik and Peter Slater have shown that they are particularly commonly heard in situations where an animal has become isolated from the other members of its pod. It is supposed that these animals effectively call out their "name" when they emit their signature-whistle and thereby broadcast their identity and give away their location. Once the separated animals are reunited signature-whistle broadcasting stops. So at least one function of the whistles is the maintenance of group cohesion.

Far more research is required before we can categorically say that dolphins refer to one another by "name" but intriguingly there are situations where one dolphin will imitate the signature-whistle of another. Are these animals calling out to them to initiate a behavioral interaction?

Summary

Communication is a sharing of information between individuals. There is no requirement for honesty in animal communication. Various channels of communication are available, but which ones are used by a particular species to transmit a particular message will depend upon the sensory system and preferences of the species concerned, upon the environment through which the signal is transmitted, and on the nature of the message to be sent.

Questions for discussion

Shrike-tanagers misuse their position of trusted sentinel in order to trick their flock mates and grab the best food available. Why do the other members of the flock tolerate this behavior?

In common with a number of other bird species blue tits exhibit a number of distinct postures that appear to serve as threats, but they have comparably few submissive signals. Why should this be the case?

Some researchers seem confident that dolphins do possess signature-whistles, names if you like. Others suggest that these calls are not referential, but that they do contain signature information – variability that does allow discrimination between individuals in much the same way that we recognize voices. Why is this distinction important?

Further reading

One of the least well understood and therefore perhaps most interesting groups of animals in terms of their communication behavior are the marine mammals. Peter Tyack and Edward Millar provide a good review of our current knowledge in the area, as well as indicating priority areas for future research in *Marine Mammal Biology: An Evolutionary Approach* (edited by A. Rus Hoelzel, 2002, Blackwell Publishing, Oxford). Among the most comprehensive reviews of animal communication are *The Evolution of Communication* by Marc Hauser (1997, MIT Press, Cambridge, MA) and *Principles of Animal Communication* by Jack Bradbury and Sandra Vehrencamp (1998, Sinauer Associates, Sunderland, MA). I would heartily recommend either of them as a very good read.

6 Foraging Behavior: Finding, Choosing, and Processing Food

A curious bird is the Pelican, its bill can hold more than its belly can!

D.L. Merritt, Nashville Banner, 1913

All animals need to find and eat food to survive. Some species feed on plants and some feed on other animals. Through foraging, easily described (but not necessarily easily explained) relationships develop between species. These may involve just two species, such as the relationship between the giant panda *Ailuropoda melanoleuca* with its specialist diet, feeding largely (but contra to popular myth not exclusively) upon the shoots, leaves and stems

Contents

Key points

• The specific principles of foraging are common to all animal species, even if the actual behaviors exhibited may be very diverse. The foraging process can be thought of as number of decisions made by the forager, such as: Which item to eat? When and where to forage? When to seek new pastures? When to share food? etc.

• Behaviors that seem at first to disadvantage a forager, like sharing, can often actually increase an individual's success. However, interference and factors such as predation risk can limit foraging opportunities.

• A range of alternative strategies can be used to maximize either the quality or quantity of food available to an animal.

• It is possible to describe the likely behavior of an animal using a mathematical model. This can help us to focus our investigations of behavior.

of bamboo. At a far larger scale the relationships might involve the complicated webs of predators and prey that may link, for example, simple planktonic plants to a top predator like the great white shark *Carcharodon carcharias* via a number of other species relationships. The various modes of foraging employed by the animals within these webs are diverse, but the basic principles involved in the process of finding, selecting and eating food are probably common to all animal species.

In ecological terms we often find it useful to divide foraging animals into herbivores (which consume plants), carnivores (which consume other animals), and omnivores (with a diet including plants and animals). Within these groups we might consider specialists (reliant upon one type of food, or perhaps even one species) and generalists (with a more varied diet). Such an approach would be of benefit if we wanted to simply describe and compare the various modes of foraging involved, for example when a bivalve mollusc filters food from a water current or an ant-lion ambushes a passing insect. However, in the context of this chapter we will place the emphasis on the specific behavioral principles involved in foraging that are general to all animals.

Foraging behavior has a range of motivations

If I were to pose the question, "how do you know when to eat?" I am fairly certain that your answer would be, "when I feel hungry". Hunger then must be considered a prime motivation for foraging behavior. However hunger should not be thought of as the only factor that can stimulate foraging or the onset of a meal. In Chapter 3 we discussed learning and saw that a very wide range of animal species can be classically conditioned to associate feeding time with an apparently abstract stimulus. Similarly humans often eat just because the clock tells them that it is mealtime regardless of their level of hunger. Social cues can be important too. There is evidence to suggest that some animals tend to take larger meals at one sitting when they share the resource with conspecifics. Presumably this behavior is advantageous, because a solitary feeder may have an opportunity to return to a food resource at a later time to take another meal, but if the food is shared there may be none left to return to. Later in this chapter we will consider other examples of the consequences of sharing and foraging behavior.

Link
Hunger has a physiological basis.
Chapter 3

Foraging decisions

Food is not uniformly distributed and not all foodstuffs are of uniform quality. Foraging animals are therefore forced to make a number of "decisions" which shape their feeding behavior. Imagine your own situation when faced with a feeling of hunger during a visit to your local town, you might "ask" yourself the following questions:

• Should I head home, or should I eat here in town?
• Do I feel like a huge meal or will a chocolate bar suffice?
• Do I eat too much chocolate – perhaps I should buy an apple?

If we ask similar questions to investigate the "decisions" made by animals in the context of our study of animal behavior, we can gain a greater insight into animal foraging.

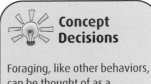

Concept Decisions

Foraging, like other behaviors, can be thought of as a decision-making process. Finding, choosing, processing and eating a food item may involve a sequence of behaviors.

Learning can shape foraging behavior

A solitary animal may be able to call upon its own experiences in making decisions such as where to forage and what to eat. For example, animals as diverse as octopus and rats can quickly learn to navigate a maze to reach a food source. Rats also exhibit an innate level of bait-shyness or neophobia, a reluctance to try new foods, which protects them from poisoning (and given their pest status, poisoning is a real possibility for rats). In fact wild brown rats (*Rattus norvegicus*) will avoid a novel food source for several days even when familiar food is scarce or lacking. Numerous experiments have repeatedly demonstrated that rats that do eat bait (and survive) can learn very quickly to avoid the poison by associating the food that was eaten with subsequent sickness. This occurs as a result of a specific form of learning termed **conditioned taste aversion** (CTA). But while some sectors of the pest control industry devote time to the development of treatments and baits that will not be avoided by animals like rats, others do precisely the opposite. For example in laboratory trials starlings (*Sturnus vulgaris*) given a choice between untreated food and food laced with garlic oil strongly avoid the latter. These birds have a good sense of smell and are thought to be reacting specifically to the sulfurous odor of the garlic. This is a good strategy because sulfurous smells often indicate the presence of toxic selenium in foodstuffs. Because starlings are agricultural pests this research could have real significance in the development of a commercial repellent for use in crop protection.

Link
Learning plays an important role in foraging behavior.
Chapter 4

I am sure that we have all at one time or another vowed never to eat a particular food (or more likely take a particular drink) again because it made us sick. This ability that we have to very quickly, and strongly, develop an aversion to a particular food is a common property of a very wide range of animal species, and one that is generally described as a conditioned taste aversion (CTA).

CTA is of course a form of classical conditioning (see Chapter 4), and a conditioned aversion can be generated experimentally if an animal is presented with food (the unconditional stimulus) that has been laced with a sickness-inducing agent (the conditional stimulus). It then associates the food with the sickness that it subsequently experiences (the conditional response).

In the laboratory E.L. Gill has demonstrated that after just one experience of being fed dead mealworms coupled with a treatment of 17α-ethinylestradiol (an effective CTA agent), rats became conditioned avoid both dead and living mealworms for up to 8 weeks (Fig. 6.1).

In conservation terms CTA has the potential to be a useful technique in those situations where a reduction in the activity of a specific predator on a specific prey population is desirable. Here the aim is not to condition the subject to avoid a taste that it associates with the foodstuff, but to condition it not to attempt to eat the food at all. There are two scenarios in which one might envisage the application of CTA protocols. Those situations in which the hunting activity of an endangered predator brings it into conflict with humans because it targets their stock animals, and those situations where predator pressure prevents the recovery of a population of an endangered prey animal. In the cases of the latter if the predator can be dissuaded from attacking the prey animal even for a short time it is possible that a positive impact on the prey animal will result. This might be particularly important if the prey animal is especially vulnerable at a particular life stage. For example, a large number of endangered bird species suffer high levels of nest predation. It has been shown experimentally that a wide range of egg predators, including crows, ravens, racoons, and mongooses, can be dissuaded from eating eggs if they are first allowed to eat one that has been injected with a CTA agent. Other experimental research has shown that racoons can be dissuaded from stealing chickens from farms, and coyotes can be dissuade from attacking lambs if they are allowed to scavenge a CTA-agent-laced chicken/lamb carcass.

As an aside, conditioning has also been successfully used to reduce human mortality as a result of tiger (*Panthera tigris*) attacks in India (but not CTA in this case!). Electrified decoys dressed as agricultural workers and fishermen were placed on tiger hunting grounds, and following the nonlethal shocking of the man-eaters attacks on humans fell from 45 to 22 per year.

Fig. 6.1 After treatment the rats exposed to a CTA agent avoided eating mealworms for some weeks. (From Cowan, D.P. *et al.* (2000) Reducing predation through conditioned taste aversion. In *Behaviour and Conservation*, ed. by Gosling, L.M. & Sutherland, W.J., pp. 281–99. Cambridge University Press, Cambridge. Reproduced with permission of Cambridge University Press.)

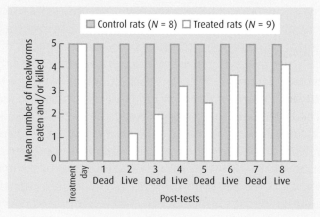

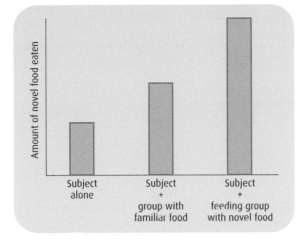

Fig. 6.2 Foraging in the company of others makes capuchin monkeys more likely to eat novel foods. (From Vissalberghi, E. & Addessi, E. (2000) Seeing group members eating a familiar food enhances the acceptance of novel foods in Capuchin monkeys. *Animal Behaviour*, **60**, 69–76. Reproduced with permission of Elsevier.)

We have already seen that social cues are important in the motivation of foraging behavior. Evidence from experiments involving a population of captive capuchin monkeys *Cebus apella* has demonstrated that they may also be important in overcoming neophobia. In their experiments Elisabetta Visalberghi and Elsa Addessi presented a number of capuchin subjects with a novel food in three different situations: as a solitary subject animal, in the presence of other capuchins that did not have access to any food, and in the presence of other capuchins that did have access to food (in this case food with which they were familiar).

The results of this experiment (Fig. 6.2) showed that the subject animals were far more likely to eat the novel food in the presence of conspecifics than they were when forced to feed alone. The effect was stronger when the other animals present were also feeding, even though they were eating different food.

Animals exploit information provided by their neighbors

Animals do not have to rely entirely upon their own food-finding abilities. Successful foragers provide information (either intentionally or unintentionally) that can be used by an individual to enhance its own foraging success. For example, it has been suggested that the breeding colonies of some bird species may have an additional function as centers for information transfer. Data to support this suggestion have been collected by several workers.

However, not all birds that breed colonially take advantage of the foraging information provided by their neighbors. As we have

Case study Osprey colonies are centers for information transfer

Throughout much of its range the osprey *Pandion haliaetus* is associated with freshwater, but in parts of North America this large fish-eating bird is coastal. Ospreys hunt fish, which in coastal waters are commonly found in shoals. This means that although it may be locally superabundant the bird's food is patchily distributed. A lucky osprey will encounter a shoal, but an unlucky bird may search without success. In this context ospreys form loose breeding colonies and successful hunters return to their nests with their prey. Given these facts we might reasonably make a number of predictions about osprey foraging behavior. Firstly, that successful birds will return repeatedly to the shoal of fish that they have found, and secondly that unsuccessful birds would benefit if they were to hunt in the area that had recently proved successful for their neighbors. This second prediction is termed the "information transfer hypothesis," the idea being that one individual gains information from cues provided by another, and then acts upon the information for its own benefit. Erick Greene tested these ideas through meticulous observations of hunting ospreys. He found that birds were significantly more likely to begin a foraging trip shortly after the return to the colony of an individual carrying fish, and that they tended to head off in the direction originally taken by the successful bird (Plate 6.1). The birds that took their cue from a successful hunter were more successful than those that did not. So it seems that ospreys do learn where to forage from their colony mates (Fig. 6.3).

In addition to the ability to recognize successful and unsuccessful hunters the ospreys appear to be able to discriminate between different prey types. From Fig. 6.3 it is clear that birds are more likely to begin fishing after the arrival of a successful bird carrying an alewife, pollock, or smelt, but that they treat a bird carrying a winter flounder in the same way that they treat an unsuccessful individual. The reason that they do this becomes clear if you consider the ecology of the four prey species. Winter flounder is a solitary animal, whereas the other three species all form shoals.

Information transfer of the kind discussed so far is probably passive. By which I mean that the bird providing the information does not necessarily intend to do so. However, Greene does report examples of active information transfer amongst the ospreys in his study population. He observed that on occasion a successful forager returning to the colony after a prolonged period during which none of the colony members had caught a fish performed an elaborate display. This consisted of a conspicuous and undulating flight into the colony accompanied by a persistent call. Moreover, such displays were most common when the prey item being carried was a

Plate 6.1 Osprey "share" information about good hunting. © P. Dunn.

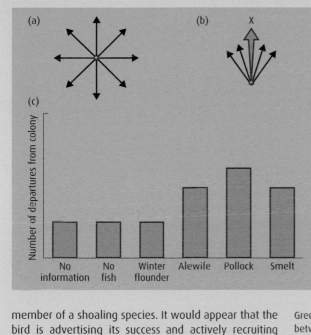

Fig. 6.3 Ospreys gain information from foraging neighbors. With no cues birds set off to forage in a variety of directions (a), but when a successful hunter returns from direction X, birds tend to set off in that direction (b). (c) Birds will take cues from a neighbor carrying an alewife, pollock, or smelt, but they are not inspired at the sight of a winter flounder. (From Green 1987. Reprinted with permission from *Nature*. Copyright (1987) Macmillan Magazines Limited.)

member of a shoaling species. It would appear that the bird is advertising its success and actively recruiting other colony members to a potential food source.

Greene, E. (1987) Individuals in an osprey colony discriminate between high and low quality information. *Nature*, 329: 239–241.

already seen in Chapter 1, black-headed gulls are colonial. Based on the information transfer hypothesis a team of researchers predicted (quite reasonably) that once one gull had located a pile of fish provided by them others would follow the discoverer to exploit the resource. The fish were dumped and a gull did gorge itself on them. It took the food to its nest and emptied its bulging crop to feed its noisy chick in full view of its neighbors. The bird returned to the fish pile several times but no other birds followed it. It would seem then that despite some seemingly very obvious cues the other gulls in the colony did not make use of the information about foraging success that was provided.

Group membership can increase an individual's foraging success

Some animals take information transfer and food sharing to a higher level and hunt in groups. The reasons for this are twofold. Firstly, being part of a group may afford some degree of protection to the individual from potential predators. Secondly, group foraging may confer an advantage upon the individual by increasing

Link
Group living means safety in numbers.
Chapter 7

Case study Predatory fish can benefit from being members of hunting schools

Peter Major has found an effect of hunting group size upon prey capture during his study of the foraging behavior of the jack *Carvax ignobilis*, a predatory fish which is often solitary, but which when feeding forms small, temporary schools. Jacks hunt the Hawaiian anchovy *Stolephorus purpureus* which is always found in large schools. Major found that jacks form feeding groups because on average a fish in a school fares far better than a solitary hunter (Fig. 6.4). However, the data also show that an individual's share of the catch depends very much upon its position in the school, and animals in fourth or fifth place are probably doing less well than a solitary hunter. So why do schools of more than three fishes ever form? Major suggests that the answer to this question rests with the observation that any fish can break off to lead an attack and thus occupy the pole position. On average, therefore, all animals probably do relatively well because they all take a turn at the higher ranked positions. He also noted that an anchovy school was always more effectively disrupted by a larger foraging group, resulting in a greater number of easier to catch stragglers.

Major, P.F. (1978) Predator–prey interactions in two schooling fishes, *Carax ignobilis* and *Stolephorus purpureus*. *Animal Behaviour*, **26**, 760–77.

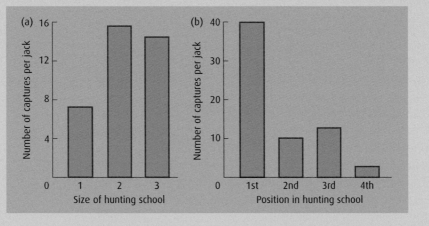

Fig. 6.4 School size is related to success when jack hunt. School size increases individual hunting success (a), but position in the group is clearly very important. (From Major 1978.)

its level of food intake. Clearly we would only expect an animal to be a member of a foraging group if it could gain more by doing so (and by sharing the resource) than it could by foraging alone (assuming all other things to be equal). In experiments under captive conditions it has been shown that flock size increases the individual fishing success of gulls, probably because fish fleeing from one bird will often swim straight into the beak of another (Fig. 6.5).

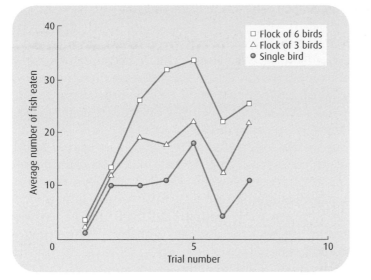

Fig. 6.5 Group size influences hunting success in black-headed gulls. (From Göttmark, F., Winkler, D. and Andersson, M. (1986) Flock-feeding on fish schools increases success in gulls. *Nature*, **319**, 589–91. Reprinted with permission from *Nature*. Copyright (1986) Macmillan Magazines Limited.)

Increased ease of prey capture cannot however explain the formation of foraging groups by herbivores. As we will see in Chapter 7, they form in part as a response to the threat of predation, but research into group foraging in the Brent goose *Branta bernicla* suggests that they may in part be due to a need to ensure foraging quality. These geese feed on the saltmarsh plants growing in their coastal wintering areas and along their migration routes. This vegetation is covered by the tide twice each day and so feeding time is at a premium. For this reason the birds make use of the foraging opportunities available at any one time and do not defend individual territories. Cropped vegetation of this kind is a renewable resource – it is diminished during a feeding bout but will regrow between bouts, and in theory the longer that it is left the more of it there will be. In practice, however, the most nutritious growth is maximal 4 days after cropping. An individual can therefore maximize its own intake by cropping a site, moving on, and then not returning to that site until a sufficient period has elapsed for its regeneration. This strategy can only work if other individuals do not also use the same area. If they do, and they revisit the next patch on your circuit the day before you do, then you will lose out and may even starve. Observations of Brent geese in Holland have shown that the birds operate like a single unit. They form a flock with a highly predictable feeding routine, returning to exactly the same patches of grazing on a 4-day cycle.

Concept Sharing resources

Sharing a foraging opportunity, via an intentional cooperation, or by synchronizing feeding activity, can actually increase individual success.

Active cooperation enhances foraging success

Some animals, such as lions and wolves, form hunting groups. In this case individuals don't simply benefit because they are hunting the same prey at the same time, they increase their hunting success as a result of active cooperation. One notable example is that of killer whales *Orcinus orca* preying upon other sea mammals. There are numerous reports in the literature of pods (the cohesive social groups in which killer whales live) working together to attack a large mammal such as a baleen whale, or an especially mobile one, such as a seal, in the open water. There are also reports of groups of killer whales swimming very close to an ice floe in order to create a swash that will wash resting seals from the ice. By working together in this way the animals probably capture prey which they would not be able to exploit alone. Killer whales hunting the population of southern sea lions *Otario flavescens* and elephant seals *Mirounga leonina* at Punta Norte, Argentina have another way to increase the foraging efficiency of the pod. At Punta Norte whales have been observed to rush from the sea, up the shallow beach, and to intentionally strand themselves to take their prey in the shallow water. Observations made by Rus Hoelzel suggest that only one member of each pod specializes in this form of hunting and that this individual shares its catch with the other pod members. It had been suggested that this is another form of cooperative hunting, and that the other pod members assist in herding prey towards an area suitable for the attack, thus earning their share. However Hoelzel's observations provide no evidence for this. He suggests that the animals in the pod are closely related to one another and that by sharing, the total foraging efficiency of the group is increased. Stranding is a behavior which requires experience and it is thought that adults teach the behavior to the younger members of the pod.

When hunting shoals of herring, humpback whales *Megaptera novaeangliae* make use of a particular cooperative behavior termed bubble-cloud feeding. This behavior involves the production of a cloud of ascending bubbles of air, each the size of a grapefruit and together having a diameter of up to 10 meters. As one whale produces the bubbles others in the group herd the herring towards them. It may be that calls made by the whales at this time aid in the coordination of this activity. The herring have a strong aversion to the bubbles and can be trapped in clumps against them as the whales lunge through the shoal to engulf their prey.

Key reference
Hoelzel, A.R. (1991) Killer whale predation on marine mammals at Punta Norte, Argentina: food sharing, provisioning and foraging strategy. *Behavioral Ecology and Sociobiology*, **29**, 197–204.

Link
Vocal communication can help individuals to coordinate their activities
Chapter 5

Deciding what to eat

Foraging animals can maximize their intake by increasing food quality or quantity

Various species of herbivorous fish play an important ecological role throughout the sublittoral, but their effect is most obvious in coral seas. It has been estimated that relatively small populations of herbivorous fishes such as the trigger and butterfly fishes may consume 50–100% of all of the algal production on a reef system, thereby preventing the smothering of the coral and thus preserving the reef. It is thought that each fish needs to consume large amounts of material to maximize its energy intake because seaweed is relatively poor in nutrients. So we might say that the behavioral solution to the fishes' problem of feeding on poor quality food is simply to eat more.

The green turtle *Chelonia mydas* faces a similar problem. Green turtles graze primarily on the sea grass *Thalassia testudinum*, which is highly productive, has a constant nutrient quality, and for which the turtles have few competitors. It would seem therefore to be an ideal foodstuff. However, *Thalassia* also has a very high cellulose content, which makes it difficult to digest. Green turtles do have an extremely efficient digestive system, but cellulose digestion is slow and so a meal has a long gut residence time. This means that whereas butterfly fishes can simply eat more seaweed to maximize their nutrient intake, turtles cannot use the same strategy. It seems likely that the turtles' grazing behavior provides a solution to their problem. Research has demonstrated that turtles consistently return to crop the same areas again and again, leaving adjacent *Thalassia* patches untouched. By doing this the turtles maximize their intake of regenerated, young shoots. These have been shown to be lower in cellulose, and up to 11% higher in protein, than shoots from the adjacent ungrazed stands.

Dietary preferences may depend upon the age or developmental stage of an individual

The marine iguana (*Amblyrhynchus cristatus*) of the Galapagos Islands are often portrayed as being subtidal grazers of macro-algae. This is not, however, an entirely accurate picture as Fritz Trillmich and his coworkers have shown. They have described

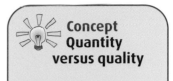

**Concept
Quantity
versus quality**

Nutrient intake can be maximized in a number of ways. An animal could increase the total amount eaten, or it could select relatively high nutrient food, i.e. it could increase quantity or quality.

**Concept
Alternative
strategies**

The fish and the turtle both face the same problem, but their solutions are very different and depend in part upon their physiology and ecology.

Within a species or population alternative behavioral strategies are common.

Investigating the underlying factors that explain this situation often leads to a greater understanding of the behavior itself.

two foraging strategies used by the iguana in their study population. Some animals do swim to feed subtidally, but others are exclusively intertidal, feeding around the low water mark on exposed seaweed.

Their observations indicate that the intertidal animals are all the smaller members of the population (less than 1.2 kg), whereas the exclusively subtidal feeders are the larger members of the population (weighing more than 1.8 kg). It would appear then that this variation in feeding behavior is related to body size. This suspicion is confirmed by the additional observation that the only animals seen to mix the two strategies are those which weigh between 1.2 and 1.8 kilograms, i.e. they are intermediate in both size and behavior. Marine iguana therefore undergo an ontogenetic change in their foraging behavior as they age and grow. The question we should ask ourselves is why do some animals eat the intertidal seaweed when others eat the subtidal plants?

Subtidal foraging has the clear advantage that it is not dependent upon either the tidal cycle or good weather. Intertidal foraging is not possible at high tide or when the sea is rough. For a reptile the constraints of the tide present difficulties to an intertidal animal which has to limit its exposure to seawater to prevent cooling (the sea around the Galapagos islands is at 14–25°C). If the tide is low shortly before dark, a feeding animal may be chilled by wave splash, but then not have sufficient daylight to reheat itself before nightfall. Subtidal foragers do not face this problem and are able to concentrate their foraging during the late morning and around noon thereby allowing time to reheat during early afternoon when solar radiation is still intense.

Why then do any animals forage intertidally? Two size-specific constraints appear to limit the size at which an animal can switch to a subtidal strategy. Firstly, smaller animals cool more rapidly than larger animals and they would therefore be restricted to shorter foraging trips (research has confirmed that the animals have a lower body temperature threshold that acts as their cue to leave the water). Secondly, iguanas are actually relatively poor swimmers and small animals are the poorest swimmers of all. The longer time taken for a small animal to reach food coupled with the fact that it has to return to reheat quickly probably restricts it to an intertidal life.

Key reference
Trillmich, K.G.K. & Trillmich, F. (1986) Foraging strategies of the marine iguana, *Amblyrhynchus cristatus*. *Behavioral Ecology and Sociobiology*, **18**, 259–66.

Case study Crabs choose their mollusc prey in an optimal fashion

Elner and Hughes have used optimal foraging theory to explain the prey choice made by shore crabs *Carcinus maenas* foraging on mussels *Mytilus edulis*. They presented hungry crabs with equal numbers of mussels, of a range of sizes, and recorded which ones the crabs chose to eat. Bigger mussels provide more calories per meal and so we might expect that the crabs would have chosen the larger bivalves in preference to the smaller ones. However, the results obtained showed that they did not (Fig. 6.6). The crabs were clearly able to open and consume mussels from the full size range, but why did they specialize in those between 2 and 2.5 centimeters long?

Optimal foraging theory makes the assumption that animals make choices that will maximize their net energy gain and in doing so take in more energy than they expend feeding. So Elner and Hughes took into account not only the calorific value of the food, but also the energy cost involved in processing and eating it. To do this they recorded the time taken by a feeding crab to open the mussel shell (an expression of the energy expenditure) in relation to the crab's calorific intake. Bigger mussels take longer to open and so the crabs may "spend" more energy to reach the food they contain. This can be expressed as an optimality model (the mathematical expression: energy divided by time or E/T), which allows the prediction of the profitablity of a mussel of a given size. Figure 6.6 shows the distribution of profitability of the range of mussels offered to the crabs in the experiment. Small mussels are easily opened but yield little, and very big mussels offer a higher yield but are difficult to crack.

The match between the most profitable mussel size predicted by the optimality model and that chosen by the crabs (compare Fig. 6.6 (a) and (b)) is a good one, and so we would say that these crabs are optimal foragers in this situation.

Elner, R.W. & Hughes, R.N. (1978) Energy maximisation in the diet of the shore crab *Carcinus maenas*. *Journal of Animal Ecology*, **47**, 103–6.

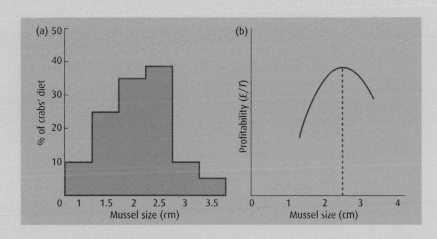

Fig. 6.6 Shore crabs are optimal foragers. Crabs preferentially select prey of a specific size (a), which allows them to maximize their energy intake (b). (From Elner & Hughes 1978.)

Optimal foraging

Animals which behave optimally balance the costs of foraging against its benefits

We have just seen that small iguana are unable to feed subtidally for a number of reasons. Another way to express their problem would be to say that the energy they can gain feeding subtidally (their net gain) is exceeded by their net energy loss (the energy used swimming and lost cooling).

Thinking about foraging behavior as a trade off between energy costs and gains is the basis of the theory of optimal foraging, which as a tool for the study of foraging has proved to be very useful.

The case study by Elner and Hughes is an excellent example of the way in which a theoretical model can help us to better understand the behaviors that we observe. Of course this model is very simplistic, it assumes that the only factors relevant to the choice relate to the prey item itself. In reality it is likely that a whole host of factors such as time of day, state of tide, risk of predation, distribution of food items, previous foraging experience, and ontogeny exert some effect. Nevertheless, so long as we recognize the limitations of the model, the results of studies such as the one we have just discussed are an invaluable guide to the underlying principles governing foraging decisions.

In an attempt to improve their applicability several workers have modified or refined basic optimality models. For example, in their excellent textbook *An Introduction to Animal Behavior*, John Krebs and Nick Davies (1993) have expanded the basic expression of profitability. They have done so in an attempt to explain the fact that even though mussels of 2–2.5 centimeters are the optimal choice for shore crabs, they do eat animals from the full size range. Their model describes an hypothetical situation in which only two prey types exist, big prey (with a profitability of $energy_1/time_1$ or E_1/T_1) and small prey (profitability E_2/T_2), and where big prey are energetically the most profitable ($E_1/T_1 > E_2/T_2$). Using this information Krebs and Davies pose the question, when it encounters a prey item should the predator eat it or continue to search for food?

As big prey are the most profitable, an animal should always eat the big prey it encounters, this is because there are only two prey types and no amount of searching will provide a more valuable meal than the one in front of it. On the other hand, if the predator encounters a small prey item it could do better if it could find a big one, so in this situation should it eat or search? It should eat if the amount of extra "profit" a big prey would yield is greater than the amount of energy it will spend in the search (i.e. if $E_2/T_2 > E_1/(S_1 + T_1)$, where S_1 is the time taken to search for new prey).

The expanded form of the simple model is useful because it allows us to make a number of predictions about the behavior of animals that we assume to forage optimally. This is a particular strength of theoretical models in that they can be used to point the way to further experimentation. This particular model can be used to make four predictions:

• A predator should either specialize on the more profitable prey or eat both (but never specialize on the less profitable prey).

Mathematical models are useful tools
Mathematical models are very useful in the study of animal behavior. They can help researchers to focus their attention on the important aspects of a particular behavior and can be used to generate predictions about the way in which we might expect an animal to behave. These predictions or hypotheses can then be tested experimentally.

Models can be constantly improved upon
It is important to remember that models are only as useful as the information they contain. For this reason basic models are often revised and improved in a way that increases their complexity and makes the assumptions they make more closely reflect nature.

Models should be tested empirically
The predictions we generate from models should be tested in the laboratory or the field. A good model will accurately describe the real behavior of an animal.

Models can be used in a range of contexts
A model need not be restricted to the study of a single behavior. By understanding the components of the model or the "currencies" of the components, it is possible to apply one model in a wide variety of contexts.

• The decision to eat big prey does not depend upon the density of small prey, whilst the decision to eat small prey does depend on the density of big prey.
• The decision to specialize depends upon the distribution of the profitable prey and not on the distribution of the less profitable prey.
• The switch from specialism to generalism should be sudden (when $E_2/T_2 > E_1/(S_1 + T_1)$).

There have been a number of empirical studies that have provided data that support these predictions. For example studies of the blue crab *Callinectes sapidus* support the first of them. The crabs show a distinct preference for the more profitable clam *Mya arenaria* when presented with a choice between it and the less profitable clam *Rangia cuneata*. However, the same crabs do not discriminate between *Mya* and the similarly profitable mussel *Ischadium recurvum*. The crabs demonstrate specialism because they were seen to excavate a bivalve, only to find that it was *Rangia* and to then discard it, uneaten, and continue to search for prey.

Evidence to support the second and third predictions comes from studies of the redshank *Tringia totanus*. These wading birds selectively hunt larger polychaetes (somehow assessing the size of their buried prey), ignoring smaller worms. The density of the small prey does not affect their decision to take large prey, but small prey are only taken when larger animals are scarce.

The fourth prediction is supported by observations of foraging plaice *Pleuronectes platessa* (termed 0-group plaice) feeding on either the siphons of the bivalve *Tellina tenuis*, or on polychaete worms. The fish switch from one prey type to the other suddenly and at different prey threshold densities depending on whether the density of *Tellina* was increasing from a low value or decreasing from a high value.

Sessile organisms can be optimal foragers

The majority of sessile filter feeders, suspension feeders, and deposit feeders do not pursue or search for their prey and so it would be meaningless to use a component like search time as a currency in an investigation of the optimality of their foraging behavior. However, this does not mean that we cannot make use of optimality theory in attempting to understand their behavior, we just need to know what component to substitute for T in the equations. They do make an investment in foraging in terms of the filtering apparatus, the rate of filtering or pumping of seawater, or the rate of sediment ingestion. However, these costs are low and evidence suggests that the energy invested in these foraging behaviors is a fraction of that invested in, for example, respiration. Similarly these organisms differ from the other consumers that we have discussed in that prey selection is unlikely to be common amongst them. Animals relying on a current to deliver food to them can probably ill afford to be too picky. In an excellent review of this topic Roger Hughes describes a number of optimality models specifically designed to incorporate currencies which are meaningful in this context, (filtering rate, particle selectivity, digestion rates, rejection rates, etc.). These models have generated testable predictions and subsequent experimental work has produced data to support the idea that without doubt the sessile benthic organisms include some of the most efficient optimal foragers.

Key reference
Hughes, R.N. (1980) Optimal foraging theory in the marine context. *Oceanographic Marine Biology Annual Review*, **18**, 423–81.

Plate 6.2 Whale sharks undertake long migrations to find profitable feeding grounds. © C. Waller.

As the resources at one location are depleted animals must decide when to move on

Almost without exception a foraging animal depletes the resource available to it. At some point the decision must be taken to move on in search of a new patch to exploit (Plate 6.2). Charvov developed a useful theoretical model in 1976, which he named the "marginal value theorem," to explore foraging decisions of this kind. He assumed that when an animal is moving between food patches it does not feed and so must have a net energy intake of zero (or to put it another way it must pay a search time cost, which we can denote with the letter S). In its initial foraging patch (patch one) the animals should feed optimally (i.e. maximize E_1/T_1) as we have already discussed. We would only expect it to move on when E_2/T_2 in the patch to which it moves would exceed $E_1/T_1 + S$. Or to put it another way, it should only move when it can gain more if it does so than it would by staying put.

However, for this situation to exist the animal would be required to know the profitabilities of all of the food patches available to it – otherwise how would it know when it would pay to move? It is unlikely that any animal could have such precise information and so it would appear that the model is unlikely to reflect reality. However, it could be possible for an animal to simply compare the patch it is in with an average patch profitability (based upon prior experience of the habitat). So Charnov has modified the model to state that an animal should leave a patch when the rate of energy gain experienced by it in patch one falls

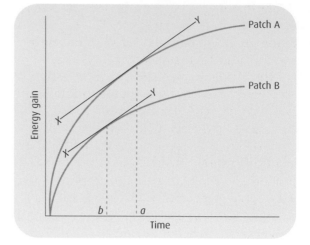

Fig. 6.7 A representation of the marginal value theorem. In this case patch A is more profitable than patch B. The lines X—Y represent the average net rate of energy in the habitat. Animals foraging in patch B reach the average net rate before those in patch A. For this reason patch B will be deserted (at time *b*) sooner than patch A (at time *a*). (From Kacelnik, A. (1984) Central place foraging in starlings (*Sturnus vulgaris*). I. Patch residence time. *Journal of Animal Ecology*, **53**, 283–99.)

to the average profitability of all of the patches in the environment. Or when rate $E_1/T_1 = E/T$ (where E/T is the average patch profitability within the habitat). If this model were to apply in nature we would presumably expect a predator to visit all of the patches in its environment, but to spend relatively little time in the less than average profitability areas and proportionally more in the better than average patches. The exact duration of a stay will depend upon both the patch yield and the distance to a new patch. This model is usually presented schematically, for example as in Fig. 6.7. As the more profitable patches are depleted the animals will move on until eventually there are none of the better than average patches left. At this point the animals should start to exploit patches that were previously ignored.

So have data been collected which support some of the predictions of the marginal value theorem? During a study of the optimality foraging of redshanks researchers divided a sandy beach used by the foraging birds into five areas. Throughout a winter season they observed the numbers and behaviors of foraging birds, and the abundance of their prey *Corophium volutator* within (and between) these patches. It was found that the birds spent most time foraging in the most productive patches but that the less productive patches were also visited (briefly) throughout the study period. Presumably these brief visits could have allowed the birds to maintain a level of information regarding the average patch profitability. As the season progressed the patches used most often by the birds were depleted of their prey and so the most profitable foraging area, and the area in which the birds hunted, changed throughout the year. These birds would seem therefore to forage in the pattern that is predicted by Charnov's model. They optimize their intake, but in doing so periodically sample all of the available patches. In another study the oystercatcher *Haematopus ostralegus* was found to forage in a broadly similar way when feeding on cockles *Cerastoderma edule*. The cockles were buried in the sediment and had a patchy distribution. As predicted the birds concentrated their feeding on a progressively smaller number of patches, which had initially been the most productive. These birds did not need to sample other foraging areas directly however. Oystercatchers break open cockles on rock anvils. The suggestion is that the birds could use the shell densities at these sites, and the densities of the anvils themselves as a patch profitability indicator as they flew over the area.

Diving behavior can be described as optimal

As was stated above, a key concept in understanding the utility of mathematical models in the study of animal behavior is the idea that a single model might apply in a very wide range of contexts. Since its conception the marginal value theorem has been applied to a great many examples of animal behavior (and not all of them related to foraging). In 1998 Paul Walton and his colleagues applied it to an investigation of the diving behavior of a number of species of subsurface pursuit hunting birds, including the guillemot *Uria aalge*. Guillemots are air breathers, and like all birds have relatively small lungs that can draw on a store of oxygen held in their extensive air sacs. This contrasts markedly with the situation in mammalian divers, like the bottle-nosed dolphin *Tursiops truncatus*, which have comparatively larger lungs, and no air sacs. Any diving animal must alternate dives with periodic returns to the surface, but presumably to maximize their hunting success they must also maximize the time that they are under the water. Or to put it another way they should maximize their dive to surface ratio. Walton and his colleagues have used the marginal value theorem to predict the optimal dive/surface patterns that one would expect guillemots to display given their particular respiratory physiology. Having made this prediction they used radiotracking equipment to follow the activity patterns of a number of birds in the wild and found that their actual dive to surface ratios agreed very closely with those that the model

Key reference
Walton, P., Ruxton, G.D. & Monaghan, P. (1998) Avian diving, respiratory physiology and the marginal value theorem. *Animal Behaviour*, **56**, 165–74

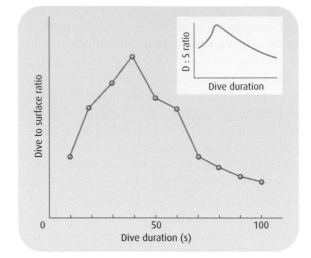

Fig. 6.8 Diving birds behave optimally. The main graph shows the pattern of dive durations recorded from a shag (*Phalacrocorax aristotelis*). Comparing the data with the optimality model (inset) shows that the bird dives in a way that will maximize its dive to surface (D : S) ratio. (From Walton, P., Ruxton, G.D. & Pitelka, F.A. (1998) Avian diving, respiratory physiology and the marginal value theorem. *Animal Behaviour*, **56**, 165–74. Reproduced with permission of Elsevier.)

predicted (Fig. 6.8). They concluded that in situations where the dives could be relatively short, the physiology of the birds allowed them to adjust their time allocation over the dive cycle in a way that would minimize their surface time and therefore maximize their hunting time.

Ideal free distributions

The distribution of foraging animals can balance the costs and benefits of group membership

The models that we have applied so far do not take into account the fact that whilst in some cases foraging alongside a conspecific may be advantageous, there are situations when it is not. We have seen that a foraging animal will deplete a food patch, but of course two animals in the same patch will often accelerate the depletion rate. In some situations the activity of one animal may directly hinder the efforts of another through competition, exclusion, or kleptoparasitism (food stealing). In other cases the effect will be indirect, the vibrations caused by one foraging wader may cause the loss of potential prey by a neighbor as the hunted invertebrates retreat into a burrow. The point is that an effective model may need to take into account factors of this nature. Models which do this, and there are a number, are explained very well by Bill Sutherland in his excellent book *From Individual Behaviour to*

Population Ecology (1996, Oxford University Press, Oxford). They stem from an original model, the Ideal Free Distribution (IFD), proposed in 1970 by Fretwell and Lucas (Fig. 6.9).

Figure 6.9 describes a situation in which a sequence of animals arrive in a habitat containing two food patches, one poor (patch B) and one good (patch A). We would expect the first immigrant to occupy patch A, this is after all the most profitable location. The second animal to arrive should also occupy patch A if it would do better in that patch (in spite of the competition/interference of the first animal) than it could do in patch B. Patch A will continue to fill in the manner described in the figure until the situation arises where an animal has more to gain by exploiting the poorer patch (B) than it would gain from the occupied patch (A). To put it another way the members of a population should distribute themselves in a way which allows each individual to maximize its own rate of feeding.

One way that feeding animals may interfere with one another is by the exclusion of subordinates from feeding areas by dominants. In this situation the IFD might predict that dominants should occupy the best patches and that subordinates might distribute themselves throughout poorer patches. It has been noted that winter oystercatcher flocks foraging on the most preferred mussel beds of an estuary contained a higher proportion of adult birds than did flocks feeding on poorer patches. It is probable that the dominant adults exclude the subordinate juveniles, and that the younger birds are distributed as per the IFD. This seems especially likely because the juveniles do in fact use the preferred beds during the summer months before the adults return on the estuary from their breeding grounds.

Foraging in a risky environment

The levels of risk experienced by a foraging animal can affect its foraging behavior

So far we have examined foraging behavior in the context of the feeding animal, the prey, and the distribution of prey. Of course foraging decisions in real life are likely to be influenced by a far greater number of factors than simply the trade-off between energy, food quality and distribution, and time. It would be impossible in a book of this kind to consider the full range of

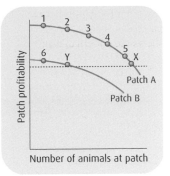

Fig. 6.9 The Ideal Free Distribution. As animals 1–5 join the foraging group they continue to gather at patch A. However the sixth individual can do better than patch A if it goes to patch B. In this example a seventh could do equally well at points X or Y.

Concept Implications of status

The dominance and subordination relationships which exist between members of a group may have an impact upon a very wide range of areas of their daily lives. Access to food, space, mating opportunities, and survival potential, for example, may in some part depend upon the individual's standing within its group.

factors involved, but in this context it is worth looking at another facet of foraging behavior, the behavior of the hunter when it is hunted itself.

To avoid predators an animal might hide or reduce its activity levels. Both strategies would be effective, but both would also be costly in that they would presumably result in reduced foraging opportunities. Another way to look at this would be to say that the animal was balancing a trade-off between feeding and, for example, hiding. The exact balance reached will depend, of course, upon the individual's motivation to feed. We might expect therefore that a hungry animal would be prepared to take more chances than one that has recently fed well.

In 1988 Carin Magnhagen carried out a number of experiments under controlled laboratory conditions to investigate the trade-off made by fish foraging in the presence of a predator. She observed the foraging behavior of two species of goby, the sand goby *Pomatoschistus minutus* and the black goby *Gobius niger* when both were hunting for buried *Corophium volutator*. To investigate the effect of predation risk half of the observations were carried out when a cod *Gadhus morhua* had been added to the tank. Her results were clear (Fig. 6.10). In all cases the presence of the predator significantly reduced both the amount of food eaten and the activity levels of the gobies. However, the magnitude of these reductions was smaller for fish that had previously been starved. This shows that the balance of the foraging–hiding trade-off is a flexible one and that the animals modify their behavior in light of their motivation to feed.

Interestingly the sand gobies maintained generally higher levels of food intake than did the black gobies throughout the experiment. Magnhagen suggests that this is a result of the fact that sand gobies are better camouflaged and therefore less conspicuous to the predator than are the black gobies in this situation. Their risk of predation is therefore likely to be lower, allowing them to be more active.

The depression of foraging activity (in relation to hunger levels) as a response to increased risk of predation appears to be a more general phenomenon. Morton and Chan (1999) have shown that in an experimental situation two scavenging gastropods, the intertidal *Nassarius festivus* and the subtidal *Nassarius siquijorensis* both move towards bait at a similar rate whether they are starved or well fed. If the animals are exposed to a predation indicator (a solution of crushed conspecific), well-fed animals are less likely to

Link
Hiding and reducing activity levels are effective antipredator behaviors.
Chapter 7

Key reference
Morton, B. & Chan, K (1999) Hunger rapidly overrides the risk of predation in the subtidal scavenger *Nassarius siquijorensis* (Gastropoda: Nasseridae): an energy budget and a comparison with the intertidal *Nassarius festivus* in Hong Kong. *Journal of Experimental Marine Biology and Ecology*, 240, 213–28.

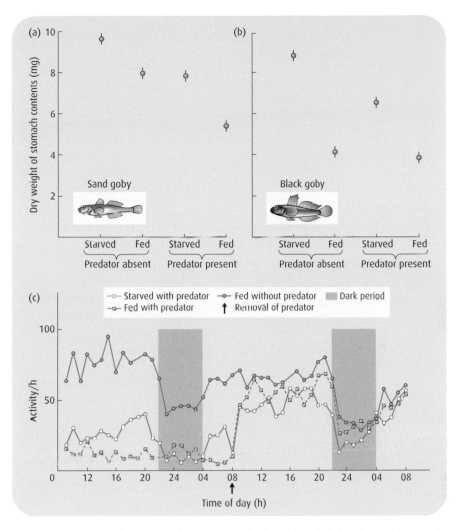

Fig. 6.10 Predators alter the hunting behavior of their prey. Sand gobies (a) and black gobies (b) have reduced hunting success in the presence of a predator, probably because they reduce the time they spend actively (and conspicuously) foraging, but if they are hungry they will take greater risks (c). (From Magnhagen, C. (1990) Conflicting demands in gobies: when to eat, reproduce, and avoid predators. In *Behavioural Ecology of Fishes*, ed. by Huntingford, F.A. & Toricelli, P., pp. 79–90. Harwood Academic Publishers, Oxford.)

forage. Starvation overrides this effect and hungry animals of both species will feed despite the risk of predation. In *N. siquijorensis* individuals, the override effect occurs between 7 and 13 days since the last meal, but in *N. festivus* it occurs later at 14–20 days. The researchers suggest that this difference stems from the fact that *N. festivus* is a smaller animal with lower energy requirements.

Summary

Throughout this chapter we have considered examples of behaviors that appear to allow an individual to maximize its foraging success. On first impressions many of these behaviors may seem to be particularly elegant immediate-term responses on the part of the animal to the situation in which it finds itself. Some of them may be just that. However, as we discussed in Chapter 1, it is important that we think of these behaviors as adaptations and remember their longer-term evolutionary context. Specifically, this chapter has considered foraging behavior as a number of decisions. It has shown that although animals may use a variety of different methods when foraging, the basic principles involved apply equally well to a diverse range of species in an equally diverse range of situations. It has been demonstrated that it is possible to think of an animal as an optimal forager. Also the utility of a range of mathematical models has been discussed and demonstrated.

Questions for discussion

Why does the information transfer hypothesis apply in the case of ospreys but not black-headed gulls?

Consider your own behavior during periods of study. Do you behave in an optimal fashion?

Is it more useful to model the behavior of an animal or to make extensive observations of it actually behaving?

Further reading

In *Unravelling Animal Behaviour* (1995, Longman Scientific and Technical, Harlow, Essex, UK), Marian Stamp Dawkins provides a very readable discussion of optimality as a behavioral concept. For a full account of foraging behavior from the point of view of animal behaviorists, ecologists, and physiologists, the volume *Diet Selection: An Interdisciplinary Approach to Foraging Behaviour* (edited by R.N. Hughes, 1993, Blackwell Science, Oxford) will provide more than enough detail for the interested reader.

7 Avoiding Predation: Staying Alive Against the Odds

Why do elephants paint their toe nails red? So they can hide in cherry trees!

Anon (but thanks to Sue Hull for reminding me)

In Chapter 6 we discussed the various behavioral adaptations that enable animals to secure the food that is essential for their survival. However, there is a flip side to this coin – remember that in many of the examples we considered animals preyed upon one another. We are used to thinking in terms of the evolutionary arms race that has resulted in the coevolution of plants and the herbivores that consume them. The plants have evolved to minimize the damage herbivores can inflict upon them and the animals in turn adapt to overcome these defences. For example, many plants possess chemicals that render them unpalatable or even poisonous. But many species of insect have evolved a way to

Contents

Key points

• Animal behavior is adapted in a wide variety of ways to avoid succumbing to predation.

• Primary defence strategies reduce the probability of an attack whilst secondary defence strategies reduce the likelihood of success of an attack.

• Behavioral adaptations are often a compromise between opposing pressures.

• There are often a number of very different hypotheses that can be generated to explain a behavior. Through careful research it is usually possible to decide which (if any) of them best explain the observed situation.

circumvent this defence strategy. Some species rely upon their physiology by sequestering or neutralizing the toxins before they can harm them, they may even use these compounds in their own defence against predators. Other animals have evolved a behavioral defence. For example, ladybird (ladybug) larvae avoid the toxins of the cucumber plant by preventing their deployment. The plant delivers the toxins to a leaf in response to grazing. The insects disarm the plant with a bite through the vein supplying the leaf they have chosen before they start to feed. The same evolutionary pressure has exerted a profound influence upon the development of the behavior of prey animals shaping their adaptive responses to predators.

Antipredator behaviors are traditionally described as either primary or secondary defence strategies. Although this distinction is often blurred and individual behaviors can sometimes fit both (or neither) category, the distinction is a useful one. Primary defence strategies can be thought of as those behaviors that reduce the probability that an individual will be attacked by a predator. Secondary defence strategies, by contrast, include those behaviors that lessen the chances of an attack being successful.

Primary defence: reducing the probability of attack

Camouflage

Concept Camouflage

Camouflage is generally thought of as a form of visual deception by which an animal can blend with its surroundings and remain undetected. Camouflaged animals are often referred to as exhibiting crypsis.

However crypsis does not have to be visual, it could be behavioral, acoustic, or involve the use of chemical odors.

Surely the best way to avoid predation is not to be detected by a predator in the first place. Many animals achieve this through camouflage. They disguise themselves as an inanimate object that the predator would not associate with food, they "disappear" against the background of their environment, or they "hide in the crowd" by closely resembling a nonprey species. To do this some animals rely solely upon their usual coloration, others alter their appearance, and some make use of the materials surrounding them (Plate 7.1).

On the rocky shore small shore crabs (*Carcinus* spp.), 1 or 2 centimeters across the carapace, come in a range of colors and patterns, some are plain and either pale or dark, whilst others have a speckled appearance. This coloration allows them to blend perfectly with the small stones and sandy gravel in which they live. A stationary animal would be very difficult for a visual predator to see. On the other hand, the larger shore crabs are almost

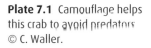

Plate 7.1 Camouflage helps this crab to avoid predators © C. Waller.

always a uniformly dark green. This change comes about, as the animal grows, to preserve its camouflage. A pale or speckled crab with a 7-centimeter carapace would be very obvious against the backdrop of a typical rocky seaweed covered shore. A dark green crab would not. The members of a number of insect groups take this process a step further. Their bodies are sculptured and molded to resemble the leaves, twigs, or flowers amongst which they live. Their camouflage matches the patterns, colors, and shapes of their environment. This disguise is made all the more perfect by the way in which the animals move. The South American dead-leaf mimicking mantis *Acanthops falcata* looks exactly like a brown, shrivelled leaf. It hangs from a twig on one stalk-like leg and twists slowly as if it had been caught by a gentle breeze. This deception provides protection from predators but in this case also improves the mantid's chances of surprising its own prey.

Counter-shading

Many species of animal, and especially birds, fish, and squid, are darkly colored on their dorsal surface but pale underneath. This typical pattern is termed counter-shading and it reduces the silhouette presented by the animal by reducing the effect of any shadows cast by the curve of the body surface due to light from above. By reducing the ease with which they are seen they avoid the attention of predators, but these species may also be able to approach potential prey more closely. The ability of the

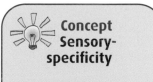

Concept Sensory-specificity

It is important to remember that patterns of camouflage may have evolved to fool a different sensory capability than our own. What seems obvious to us may be cryptic when viewed by the predator and vice versa.

cephalopods to alter their body color makes them masters of counter-shading and a number of species including the squid *Sepia officinalis* and the common octopus *Octopus vulgaris* have a counter-shading reflex. This reflex functions such that whenever the animal's body tilts or rolls by more than 90° the area of shading "moves" around the body to compensate.

Altering one's appearance

This ability of the cephalopods to alter their color pattern allows them to match their background almost perfectly. By altering their body posture (to be long and thin like a sea-grass, to hang motionless amongst floating weeds, or to take the shape of a coral head) the cephalopods have become the undisputed masters of camouflage. Why then do some species apparently ignore this ability, and sport bright blue circles or zebra stripes? In part the answer to this is that in some circumstances being bold and standing out may actually be an effective way of staying hidden. The effect of these patterns is to break up the outline of the animal against a complex background of colors and shadows. This disruptive coloration can be highly effective when coupled with the cephalopod ability to change shape rapidly. (Bold patterns may also communicate unpalatability or potential danger to predators, as we will see later in this chapter.)

Roger Hanlon and his colleagues have described what they refer to as "a particularly elegant behavior" (and I certainly agree that it is) performed by the Indo-Pacific octopus *Octopus cyanea*. They have called it "the moving rock strategy". During observations of foraging octopuses, which they considered to be at risk of predation, they noted that when crossing open areas of sand the animals often took the shape and coloration of a rock typical of that area of seabed. The animal then used only the tips of its arms to move slowly and with apparent stealth across the sand. In fact the motion was so gradual, and the disguise so good, that the researchers felt that the octopus would be indistinguishable from regular stationary rocks unless an observer paid particular attention. They also suggest that this behavior is particularly impressive in that it requires an assessment be made of the immediate habitat, and a decision be made to mimic exactly (in color, texture, and shape) local rock structures. Further, because these rocks are often some distance from the octopus, the animal must ignore visual input from the immediate environment. It must also simultaneously process information about current action, wave motion, and the dappling of the light on the seabed, correcting its disguise to take them into account.

There is another facet to this phenomenon. In their study of *O. cyanea* Hanlon's team found that foraging individuals were highly cryptic 54% of the time, moderately cryptic 24% of the time, and conspicuous 22% of the time at a study site in Tahiti. At another site in Palau the values were 31%, 19%, and 50%, respectively. Clearly these animals are not relying entirely on camouflage and the moving rock strategy as a defence. Throughout the study the octopuses were observed to change their appearance with amazing rapidity, on average once every 30 seconds or so. As we would expect this was often to maintain crypsis against a changing backdrop, but even in the open or crossing an expanse of uniform sediment rapid change in phenotype, or **transient individual polymorphism**, was common. In these situations the forms taken were often highly conspicuous and on occasion the researchers felt that the animals more closely resembled a fish or a confusion of bold patterns than they did an octopus. It is likely that this behavior is an antipredator adaptation that results in any one octopus phenotype being rare. As predators are known to develop a positive search image, i.e. to preferentially hunt familiar prey, this apparent rarity may help to protect the octopus.

Not all species can rely on their own camouflage. Many make use of the materials around them. Desert species will burrow into sand, hide under stones, or lurk in crevices so that they can avoid detection. Despite its ability to blend with the background the squid *S. officianalis* adopts a similar strategy, burying itself so that only its eyes protrude above the surface. The small sepiolid *Euprymna scolopes* has a specialized epidermis to which sand adheres. It uses this for camouflage, albeit primarily for hunting rather than protection. This coat can be quickly shed if the animal needs to take flight and the falling coat of sand may serve to confuse the attacker.

Among the masters of this kind of deception are those species of spider crab commonly referred to as decorators. The shells of these crabs are covered with tiny hooks to which they attach a blanket of material collected from their local environment. This can include pieces of seaweed and other detritus, but it can also include other animals such as anemones or sponges. The crab spends the daylight hours crouching over its folded legs and claws on the seabed and must look like a small, encrusted stone.

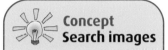

**Concept
Search images**

Predators develop search images. They learn what food looks like, and once they have done so experienced individuals are able to recognize potential prey more quickly than their naïve counterparts. So prey animals that can alter their phenotype and hamper the development of a search image are at an advantage.

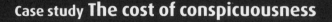
Through an experiment that demonstrates dramatically the costs of being conspicuous, researchers have shown the value of cryptic plumage in the great tit (*Parus major*). The usual coloration of these birds is a combination of yellows, grays and greens with extensive patches of black and white. This may sound conspicuous, but in fact it camouflages the birds very well against the mottled backdrop of their woodland habitat. While the

birds were still confined to their nest boxes and dependent upon their parents, the researchers fitted them with metal leg-rings to enable individual recognition. They then experimentally manipulated their plumage. Some individuals had the white feathers of their cheeks, tails and wings painted red (making them more conspicuous), and in some birds the same feathers were painted yellow (making them less conspicuous). The birds were allowed to fledge naturally and approximately 2 weeks after they had left the nest the researchers attempted to measure the predation rate suffered by them. Measuring predation directly is difficult in a highly mobile species, but in this case it was possible because the main predator of great tits is the sparrowhawk. In fact during the first weeks of life young tits are particularly vulnerable to hawk predation and the sparrowhawks time their own breeding activity to coincide with the emergence of this bonanza food resource. Twenty-seven pairs of hawks nested within the study area and the areas around their nests, and the nests themselves were searched using a metal detector to locate tit leg rings. The results of the search indicated that red painted birds were 38% more likely to be eaten than were the yellow birds and so it would appear that conspicuousness carries a cost (Fig. 7.1).

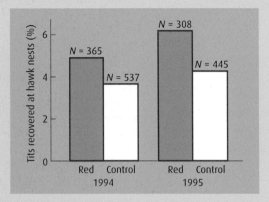

Fig. 7.1 The percentages of red-painted and yellow-painted (control) great tits recovered from the nests of sparrowhawks during the two years of the study. *N* indicates the total sample size, i.e. the number of tits that were painted. (From Götmark & Olsson 1997. Reproduced with permission of Elsevier.)

Götmark, F. & Olsson, J. (1997) Artificial color mutation: do red-painted great tits experience increased or decreased predation? *Animal Behaviour*, **53**, 83–91.

At night when visual hunters are less active, and its outline will pose less of a risk, the crab moves around to feed, carrying its protective coat with it.

Predator distraction displays

Although remaining inconspicuous clearly has antipredator benefits, there are occasions when advertising your presence is an advantage too. We will consider a number of examples of this phenomenon later in this chapter, but here I want to discuss a set of antipredator behaviors employed by the plovers and other ground-nesting birds to protect their vulnerable eggs and chicks. Plover nests are simple scrapes in the ground and they are most commonly found in very open situations such as beaches or open fields. A vigilant plover will detect the approach of a predator, typically a fox or a mustelid, when it is still some distance from the nest and offers little immediate threat.

When it has spotted the approaching danger the plover leaves the nest site and moves quietly away from it. Once it is a safe distance from its nest the bird employs one of four strategies. If it has moved a sufficient distance the bird will pretend to be incubating a clutch of eggs and allow itself to be flushed from the false nest. The predator will search the area diligently, but of course it will fail to find anything worth eating. If the parent bird has moved towards the predator and is in long grass it will turn suddenly and scurry noisily away, crouching and squeaking to imitate a fleeing rodent. The predator will give chase and will be lead away from the nest. Alternatively the bird may run towards the predator calling loudly, only to change direction at the last minute and run away, with the predator once again following it.

Perhaps the most well known of the plover strategies is the **broken wing display**, and I can vouch for its effectiveness having fallen for it myself on several occasions! During the display the bird flees along an irregular route (but of course always away from the nest) pretending to be injured. It will hold one or both wings at an awkward angle and will flap them noisily and ineffectually. It will make shrill calls and may even attempt, unsuccessfully, to take to the air. What predator could fail to chase such a helpless animal? When the predator has been lured a sufficient distance the plover takes to the air and will return to its nest via a circuitous route.

An important point to make is that not only do the plovers appear to assess the likelihood of success of each of the four strategies, and then adopt the most suitable one, they are also able to distinguish an actively hunting predator from a harmless grazer sharing their field, an ability not restricted to the adult birds. The chicks of many ground-nesting birds are known to exhibit tonic immobility when caught in the jaws of the predator. Put simply this means that they feign death. As the birds are seldom alone the predator will drop the "dead" chick and attempt to catch (and kill) another of its nest mates. At this point the "dead" chick will miraculously resurrect itself and flee the scene. Laboratory

Link
Communication serves an antipredator function.
Chapter 5

experiments carried out by James and Carol Gould have shown that this death-feigning behavior can be stimulated in a chick by trapping it and exposing it to a pair of eyes (marbles on sticks are a sufficient approximation). As long as the "eyes" remain present the chick will play dead. Every 30 seconds or so the chick will very slowly open one eye and look for the predator. If the eyes are still there it doesn't move and slowly closes its eye again, but if they are gone it will run for cover. Two predator eyes are essential for the stimulation of this behavior, presumably because they mimic the forward-looking eyes of a predator. A single eye is insufficient and may resemble more closely the eye of a herbivore, mounted as they are on the sides of the animal's head. After all the chick would gain no advantage if it feigned death every time a sheep or cow passed close to its nest.

Mimicry

Link
Learning can play an important role in antipredation behavior.
Chapter 4

Many animals are unpalatable, or have the potential to injure any predator foolish enough to attack or eat them. In the same way that they learn how to recognize food, predators also learn how to avoid these dangerous nonfood animals. Clearly it would be advantageous to all concerned if a predator could identify the animals to avoid, and so prey species advertise their unpalatability, the fact that they are poisonous, or their ability to fight back. In such a situation standing out is an advantage and so bright and contrasting patterns of reds, yellows, and black are common. Other species perform eye-catching displays or emit particular sounds that serve the same purpose. Traditionally two forms of mimicry are distinguished. When a group of species closely resemble one another and all are honestly advertising the fact that they are poisonous to eat, then we have an example of mullerian mimicry. Here each of the species involved benefits from the fact that they all look, sound or act alike. For example, a bird might learn to avoid eating members of the genus *Vespula* (the wasps) after an encounter with members of just one vespulid species because it associates the black and yellow coloration common to them all as an advertisement of their unpalatability. This mimicry is not confined to this single genus because many unpalatable insects have similar warning colors. Thus the bird may avoid not only the vespulids, but also members of the genus *Bembex* (the digger wasps), and even the physically very different caterpillars of the moths of the genus *Callimorpha*. The species mentioned thus far are all honestly signaling their unpalatibility, but there are mimics that are less honest. For example, the same black and yellow pattern will also protect the edible hoverflies of the genus *Syrphus*. Here the edible mimic is taking advantage of the fact that its model is inedible, an example of batesian mimicry.

Living in groups: antipredator cooperation

In Chapter 6 we considered the benefits of group living in terms of the increased success a foraging animal may have, and in doing so hinted at some of the antipredator benefits being a member of a group may provide their prey. One clear benefit is the dilution effect, the idea that the larger the group the lower the probability that a single individual will be the target of a predator. This effect probably explains why female eider *Somateria mollisima* crèche their young into large flocks during their first vulnerable days at sea. During this period the young ducks are easily picked up and swallowed by large gulls. By "looking after" the young of several females an eider will significantly reduce the probability that her own young will be the ones taken. The dilution effect is certainly one of the reasons that many species reproduce colonially, and the reason that a great majority of invertebrate species reproduce synchronously. By doing so they simply overwhelm the predators' capacity to eat (Plate 7.2).

During a study of the interaction between hunting sparrow-hawks *Accipiter nisus* and peregrine falcons *Falco peregrinus* and

Link
Cooperation during breeding, or breeding colonially, can enhance fitness.
Chapter 8

Plate 7.2 These colonial guillemots synchronize their breeding efforts. © G. Scott.

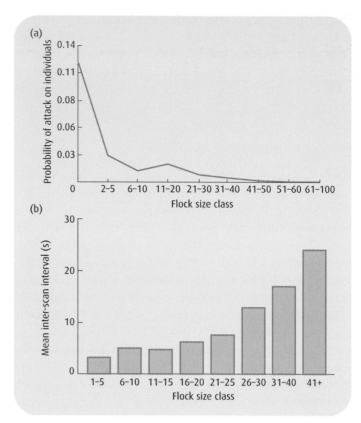

Fig. 7.2 As flock size increases the probability that any one individual will be attacked decreases (a), and the time that each individual devotes to scanning for predators decreases (b). (From Cresswell, W. (1994) Flocking is an effective anti-predator strategy in redshanks, *Tringia totanus. Animal Behaviour*, **47**, 433–42. Reproduced with permission of Elsevier.)

their redshank prey, Will Cresswell proved the applicability of the dilution effect (Fig. 7.2). His data also demonstrated the importance of two other antipredator benefits of group living, confusion and vigilance. We will return to behaviors designed to confuse a predator later in this chapter. However, because the confusion effect is in this case a characteristic of group membership we will consider it now. The basic hypothesis is that when faced with a large and mobile group of potential targets an attacking predator may become confused and unable to "lock-on" to any one of them, thereby failing to make a kill. As actual kills are comparatively rare events this effect is very difficult to study directly. Cresswell was able to investigate it indirectly because the predators he was observing have two distinct methods of attack. Some of the time they make a surprise attack, targeting an individual redshank. Some of the time they make a nonsurprise attack, apparently picking a target "on the spot." Based on this

Plate 7.3 These grazing geese share vigilance. © G. Scott.

dichotomy Cresswell suggested that nonsurprise attacks should be more vulnerable to confusion than should surprise attacks. He therefore constructed the hypothesis that the predators should target nonsurprise attacks at solitary birds (when there can be no confusion) and target surprise attacks at flocks of birds to reduce confusion. His data supported his argument and proved his hypothesis to be correct.

There is another reason that prey animals form foraging groups, and that is increased vigilance (Plate 7.3). An individual redshank is faced with a choice when feeding. It could spend all of its time being vigilant, looking out for approaching predators. If it did so it would certainly significantly reduce the chance that it would be taken by surprise, but it would also starve. A bird with its head in the air scanning for predators cannot at the same time have its head down searching for food. In reality of course an individual balances the two behaviors in accordance with the situation in which it finds itself, and as a member of a group it can shift the balance towards feeding. As Fig. 7.2 shows, the bigger the flock of birds, the less time an individual bird devotes to vigilance. This is possible because the presence of many sets of eyes in the flock effectively means that there is always somebody on the look out. There are further flexibilities built into this system because all redshanks (regardless of flock size) showed increased vigilance shortly after a predator had passed through the area.

In some circumstances the relationship between group size and vigilance is not so clear cut as one might expect. Neil Metcalfe made observations of a range of wading birds feeding in multispecies flocks on the rocky shore. He noted that some individuals had a higher level of vigilance than expected given the size of the flocks that he felt that they were part of. The reasons for this were twofold. Firstly, local topography was important. Although the birds on the beach may have appeared to all be part of one flock from the perspective of a human observer, the irregular nature of the shore meant that within an area individual birds were often invisible to one another for some of the time. During these periods it was apparent that birds did not trust individuals that they could not see to be vigilant on their behalf. The second reason that some birds were more or less vigilant than might be expected was because the study animals may have had their own views on who was and was not part of their flock. Metcalfe found that each species tended to share vigilance only with those species of similar size and ecology to themselves. Presumably, therefore, they only trusted birds facing similar risks to themselves.

The benefits of being a member of a group might not be shared equally between all group members. In Chapter 6 I mentioned that hunting jack preferentially targeted stragglers on the periphery of an anchovy school. Similarly crows or large gulls predating the nests in a black-headed gull colony are far more likely to attack nests on the edge of the colony, and so central sites are much sought after and strongly contested. This observation has been formalized as the "selfish herd effect" which states that it should always pay an individual to be at the center of a group, and to use its group mates as shield against attack. I have already suggested that anchovies at the edge of the school are more at risk than those at the center, but it should be remembered that if there are no stragglers the jack are likely to strike at the center in an effort to scatter the fish (Plate 7.4). This suggests that although the selfish herd model may apply in some situations, in others it may not.

The model certainly does appear to apply when we consider the behavior of Adélie penguins *Pygoscelis adeliae* when faced with the threat of predation by leopard seals *Hydrurga leptonyx*. The seals patrol inshore waters close to penguin breeding colonies, waiting for the birds to enter the sea in search of food. There are few places at which the birds can enter the water and there is a very high probability that there is a seal at each of them. Therefore the best strategy for the penguins would be to enter the water as a group, confuse the predator, and rely on the dilution effect for protection until they are a sufficient distance from land

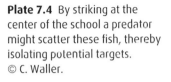

Plate 7.4 By striking at the center of the school a predator might scatter these fish, thereby isolating potential targets. © C. Waller.

to be safe. After all a seal can only kill one penguin at a time. In accordance with the selfish herd principle the birds jostle one another not to be first or last into the water and in the end synchronize their "jump-off." There are anecdotes about birds being pushed into the sea by the group to test for the presence of a seal, but this probably happens accidentally as a result of the jostling process and no premeditation has yet been proven!

As I have already stated a particular behavior is often about trading off one set of priorities against another. The selfish herd can only apply when the benefits of being at the group's core outweigh those of being at the periphery. Jeffery Black and coworkers have shown that for foraging barnacle geese *Branta leucopsis* feeding on coastal fields the best position is often at the flock edge. Birds on the edge do spend more time being vigilant to compensate for their exposed position, but they benefit greatly by gaining access to better grazing. This is because flocks move as they feed and edge birds reach new food all of the time, whereas those in the body of the flock pick over the areas where the edge birds have already grazed.

Is there an optimum group size?
The benefits of being a member of a foraging group in terms of protection from attack by hawks are clear in the case of wading birds on exposed sand or mud flats. However, as any bird watcher will tell you, it is still common to see members of a species usually found in flocks foraging as individuals. These birds may even defend their foraging areas against intrusion by conspecifics, and

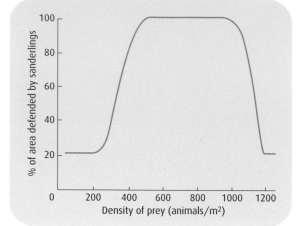

Fig. 7.3 Sanderling do not defend feeding territories when prey densities are very low or very high. (From Myers, J.P., Connors, P.G. & Pitelka, F.A. (1979) Territory size in wintering sanderlings: the effects of prey abundance and intruder density. *Auk*, **96**, 551–61.)

so are clearly not about to become members of a flock. Myers and coworkers investigated the foraging and flocking behavior of sanderling *Calidris alba*, a small wading bird, when it was hunting the isopod *Excirolana linguifrons* on sand. Some of the time the birds defended territories, but some of the time they did not. The main factor controlling their flocking behavior seemed to be the density of their prey (Fig. 7.3).

When prey were scarce the birds did not defend – it would be impossible to defend a large enough patch of beach for them to satisfy their needs. When prey were very abundant the birds did not defend – the food bonanza meant that there was enough to go around and in any case there were so many birds exploiting it that defence would have been impracticable. At median densities though the birds did defend their resource. Here the benefits of investing in defence and excluding others presumably outweigh the benefits of flocking, and it is possible that the birds are distributed in a way which could be explained through the application of the IFD (Plate 7.5).

Models to predict optimum group size in situations such as the one that we have just discussed suggest that group size should be thought of as a trade-off between the costs of being alone (more time being vigilant, etc.), the costs of group life (more competition, interference from others, etc.), and the time allowed for feeding (Fig. 7.4). The optimal flock size would be one that allowed an individual to maximize its food intake.

In reality a group is unlikely to be of optimum size in a system where immigration into and emigration out of an area is possible.

Link

Models such as the ideal free distribution (IFD) introduced in Chapter 6 can be applied in an often wide range of situations.
Chapter 6

Plate 7.5 The size of this turnstone foraging flock will be determined by a range of factors. © G. Scott.

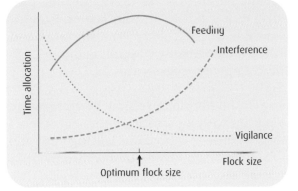

Fig. 7.4 A simplified model of optimum flock size. The optimum size is that which maximizes foraging success. You should note that this will not necessarily also be the size that minimizes interference and maximizes vigilance. (From Krebs, J.R. & Davies, N.B. (1993) *An Introduction to Behavioural Ecology*. Blackwell Science, Oxford. Reproduced with permission of Blackwell Publishing Ltd.)

As new individuals arrive the situation described in Fig. 7.4 suggests that they would be better off joining an already optimal or near optimal group than remaining alone. Eventually the group will become too large and break up, but it is unlikely that the addition of just one or two animals will have this effect. So rather than observe an optimal group size we are more likely to see groups of a "stable" size range around the optimum.

Mobbing

Often when a predator is detected in a particular area the local prey population will mob it, even if it is not actively hunting. Mobbing behavior is common in a range of vertebrate and invertebrate species, but it has most commonly been studied, and is therefore perhaps best understood, in the case of small passerine birds reacting to the proximity of avain predators. When it occurs mobbing usually takes the form of a group of individuals gathering around the predator, vocalizing constantly, and performing very conspicuous visual displays or making attack flights towards it. These flights are not always mock attacks and the predator is occasionally struck physically. The net effect of mobbing is to impair the efficiency of the predator by reducing its success rate, and perhaps ultimately to drive it away. Of course the behavior probably also serves to let it know that it has been detected and that it is unlikely to have the element of surprise that may be essential to its hunting strategy.

Hans Kruuk has demonstrated the success of mobbing as an antipredator strategy experimentally. Having noted that black-headed gulls will mob predatory crows attempting to steal their eggs, he created a number of fake nests at regular intervals from the outer edges of a gull colony to its center. Eggs in fake nests placed just outside of the colony were predated and the gulls rarely mobbed crows in this area. Mobbing increased towards the center of the colony and eggs in this area were rarely found or eaten (Fig. 7.5).

So mobbing is a highly effective strategy and few predators are prepared to put up with a persistent mob, but surely it must be risky? Coming so close to a predator is dangerous and there are numerous examples of a mobbing bird being grabbed and eaten. However, by mobbing in a group, individuals gain the advantage

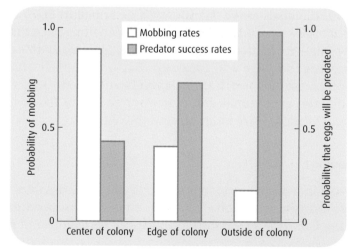

Fig. 7.5 Inside the gull colony mobbing rates are high and predator success rates are low. As nests are placed away from the colony center or outside of it they are less likely to be defended and more likely to be predated. (Data from Kruuk, H. (1964) Predators and anti-predator behaviour of the black-headed gull *Larus ridibundus. Behaviour Supplements*, **11**, 1–129.)

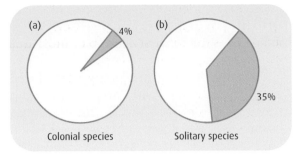

Fig. 7.6 Colonial species of swallow take fewer risks (a) than do solitary species (b) when mobbing predators. In both cases the shaded area of the diagram refers to the percentage of high-risk mobbing undertaken. (Data from Brown, C. & Hoogland, J.L. (1986) Risk in mobbing for solitary and colonial swallows. *Animal Behaviour*, **34**, 1319–23.)

of a lower personal probability of being grabbed. In fact comparisons of the mobbing behavior of group-living and solitary swallow species have demonstrated that solitary mobbers are forced to take bigger risks for mobbing to be effective. They approach their target more closely than do mobbers in a group, presumably because the increased size of the group has an increased effect on the predator. So here is another antipredator advantage associated with group living (Fig. 7.6).

Secondary defence: reducing the success of the attacker

Once detected, targeted or attacked by a predator an animal must execute some appropriate behavior in an attempt to preserve its life. Given sufficient warning, a speed advantage, or a convenient refuge, it may be enough to flee. This strategy works very well for marine molluscs such as the scallop *Chlamys opercularis* which "flies" away from an attacking starfish by clapping closed the two valves of its shell to express the water between them in an explosive propellant jet.

However, not all animals can out-run or out-maneuver a predator and so there are a range of other options available to most prey species.

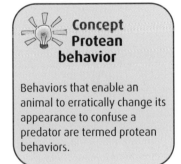

Concept
Protean behavior

Behaviors that enable an animal to erratically change its appearance to confuse a predator are termed protean behaviors.

Misdirecting the attack

Behaviors that direct the attacking predator away from vital body organs and towards expendable but conspicuous parts of the body are common and must have evolved a number of times in a very broad range of taxa. For example, the marine bivalve

The **blanch-ink-jet maneuver** is a classic example of a typical cephalopod protean behavior. It is a relatively fixed sequence of behaviors which when performed offer the animal an excellent chance of escape. The maneuver takes advantage of the cephalopod's ability to squirt ink, change its body color rapidly, and move quickly. The sequence begins when an initially dark colored octopus or squid is threatened by a predator. It responds initially by blanching, i.e. it changes its body color extremely quickly from dark to very pale. At the same time it compresses its ink sac to squirt a pseudo-morph into the water. The ink of the pseudomorph has

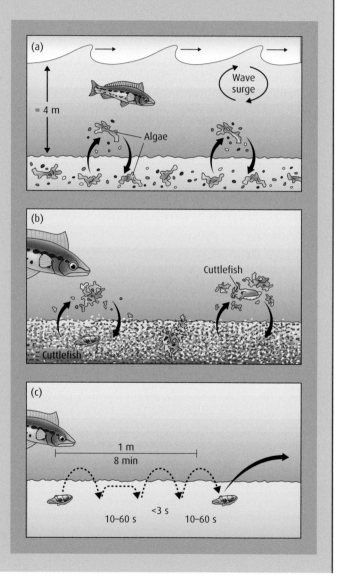

Fig. 7.7 (a–c) "Clandestine escape" by young cuttlefishes, *Sepia officinalis*. See text for an explanatory description. (From Hanlon & Messenger 1996, with permission from Cambridge University Press, based on Hanlon, R.T. & Messenger, J.B. (1988) Adaptive colouration in young cuttlefish (*Sepia officinalis* L.): the morphology and development of body patterns and their relation to behaviour. *Philosophical Transactions of the Royal Society, Series B*, **320**, 437–87.)

a high mucus content and as such "hangs" in the water holding its shape. Long thin cephalpods produce long thin pseudomorphs and rounder animals produce rounder ones. As this dark, prey-shaped object appears at the same time that the real prey animal in effect disappears due to blanching, the predator's attention is held by the pseudomorph. Now the sequence of behaviors is completed and the cephalopod really does disappear as it jets to safety.

Young cuttlefishes take advantage of local wave and sediment conditions when they perform another protean behavior, the **clandestine escape** (Fig. 7.7). As wave surge passes over the shallow seabed, detrital material (pieces of seaweed, etc.) is periodically lifted up into the water, moved a few centimeters in the direction of wave motion, and then redeposited. As successive waves pass material "leap-frogs" across the seabed. When a young cuttlefish encounters a predator that it is unlikely to be able to out-swim it joins this material. Changing color and shape to blend with the seabed it waits for the next wave to pass. Then as the detritus is lifted into the water the animal swims with it taking the shape, color and manner of movement of the material around it. Then as the wave passes and the detritus settles the animal returns to the seabed readopting its initial disguise. In this way the cuttlefish leap-frogs to safety through a series of rapid and highly coordinated phenotypic changes.

Hanlon, R.T. & Messenger, J.B. (1996) *Cephalopod Behaviour.* Cambridge University Press, Cambridge.

molluscs of the genus *Lima* are seemingly attractive prey for a range of predatory fish. These animals appear to court danger because they bear very conspicuous tentacular structures that one might assume would tempt any passing fish to nibble them. But any fish that takes the bait is in for a nasty shock. The tentacles are automotized or shed so the fish picks up far less than it expected to. They are also highly unpalatable, packed as they are with distasteful chemicals. The unfortunate predator quickly spits them out and leaves the mollusc unharmed.

Of course the most familiar example of this kind of behavior concerns the tail shedding or automoty of various species of lizards. The tails of these animals are often brightly colored, or have a pattern that contrasts with that of the main body of the animal. To make them even more conspicuous they are twitched and flicked when the lizard is threatened by a predator. A number of studies have shown that the net effect is that the lizard's tail is the target of a predator's initial strike far more often than the vulnerable head or body. When it is attacked the tail is automotized and then continues to writhe and thrash on the ground in front of the predator. The lizard of course makes a dash for escape at this point. Researchers have shown that it gains vital extra seconds because it takes a predator significantly longer to subdue the shed tail than it would have done to subdue the lizard to which it belonged.

Alarm calls

As we discussed in Chapter 5, individuals of a range of species are known to raise the alarm when they detect a predator. Some species use a visual signal, others release an alarm chemical, many make a call. We have seen previously that the nature of the signal will depend upon the environment through which it is transmitted and the sensory capabilities of the animals involved. In the case of wading birds the signal is a shrill call. The function of this vocalization is probably twofold. Firstly to "warn" other group members of their imminent danger, and secondly to let the predator know that it has lost the element of surprise. Even when they do have surprise on their side predators are less successful than we might at first assume. Without it, it may not be worth their while even to attempt an attack. Investigating this possibility is of course very difficult. We can be sure when a predator does attack but can we ever really know that it has "decided" not to? We can however investigate the effect of alarm calling on the group and upon the caller itself, and this is exactly what Will Cresswell has done as part of his study of the redshank that we discussed earlier in this chapter.

He found that although not all escape flights were accompanied by an alarm call, the redshank did call more often when attacked by a raptor than when apparently "escaping" from no discernable threat. Further, the calls were most commonly given when the birds were feeding in a visually obstructive habitat, i.e. in one in which they would be invisible to many conspecifics. In open ground they tended not to call, instead they engaged in a fast and directed escape flight. The effect on conspecifics of both behaviors was the same, they all joined the escape. This suggests that the function of the call is to elicit a "coordinated" escape, particularly when birds are unable to see one another. This is an important point. It could be argued that a caller is acting out of altruism, betraying its own location to the predator for the good of its neighbors. The evidence suggests a far more selfish motive. By sharing the information and causing the mass escape an individual can benefit from the dilution and confusion effects. In fact Cresswell found no evidence that callers were placed at a disadvantage and showed that the birds most likely to be the target of the attack were late flyers and noncallers.

Will Cresswell made at least one other important observation during his study. Remember that the redshanks he observed were

Link
Remember that alarm calls may have many possible functions; alerting conspecifics to danger, communicating with the predator, and attracting the predators of the caller's attacker.
Chapter 5

Key reference
Cresswell, W. (1994) The function of alarm calls in redshanks, *Tringa totanus*. *Animal Behaviour*, **47**, 736–8.

When fleeing from a hunting cheetah (*Acinonyx jubatus*), a Thompson's gazelle (*Gazella thompsoni*) will often interrupt its escape flight with a number of high vertical leaps with its legs held straight and stiff beneath its body. This behavior is termed "stotting", and in the face of it seems to be a strange thing to do when fleeing for one's life. A number of hypotheses to explain stotting behavior have been proposed. Some of them suggest that it maintains prey group cohesion or that it enables the gazelle to gain a better view of the hunter. Others attribute to it an "alarm-call" function, suggesting that it serves to warn conspecifics of the present threat.

In his excellent review paper on stotting Tim Caro discusses a total of 11 hypotheses including those that I have mentioned. Then in a very full and rigorous empirical field study he tests all of the various hypotheses before deciding upon the most likely explanation for the behavior. Here we will discuss just three of these potential explanations in more detail, all of them focusing on the relationship between the predator and prey animals:

Hypothesis 1: The gazelle invites an attack

It has been suggested that by persuading a cheetah to start a premature attack the "confident" gazelle can "get it over with" so to speak, and having done so return to the important business of feeding.

Hypothesis 2: The gazelle startles its attacker

The sudden leap exposes the white rump of the gazelle. This sudden flash startles and confuses the cheetah giving the gazelle the opportunity to leap to safety.

Hypothesis 3: The gazelle deters its pursuer

By stotting the gazelle communicates to the cheetah that its chances of a successful chase are slim and that it should abandon the hunt.

So which of the three hypotheses best explains the behavior given the data presented in the figures above? The **invitation to attack hypothesis** seems unlikely on theoretical grounds. Surely a predator repeatedly falling for the "trick" would come to associate stotting with a lack of success and learn not to chase a stotting individual rather than be encouraged to do so? The data

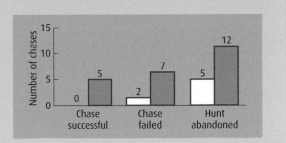

Fig. 7.8 The outcomes of 31 cheetah hunts when their gazelle quarry did (white bars) or did not (blue bars) stott. (Data from Caro 1986. Reproduced with permission of Elsevier.)

presented in Fig. 7.8 do not support this hypothesis either. Whilst the hypothesis predicts that cheetahs should chase more stotters than nonstotters, the opposite is clearly the case.

However, nor do the data support the **startle hypothesis** from which we would predict that stotting should occur at all distances from the predator, but that it should be most common when the hunter is very close and the need to startle is highest. Figure 7.9 shows that this is not the case.

By contrast, we would predict from the **pursuit deterrence** hypothesis that stotting should begin at a "safe" distance, i.e. when there is little chance of immediate capture and the gazelle is "confident" of escape. We would also predict that nonstotters should be killed more often that stotters, and that stotting animals should be chased less than nonstotters. Tim Caro's data do support these predictions and so it would seem that stotting (and a range of similar behaviors performed in the same context by a wide range of animals) do serve as pursuit deterrent signals.

Caro, T.M. (1986) The functions of stotting: a review of the hypotheses. *Animal Behaviour*, **34**, 649–62.

Caro, T.M. (1986) The functions of stotting in Thompson's gazelles: some tests of the predictions. *Animal Behaviour*, **34**, 663–84.

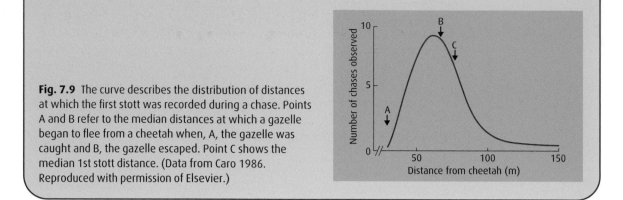

Fig. 7.9 The curve describes the distribution of distances at which the first stott was recorded during a chase. Points A and B refer to the median distances at which a gazelle began to flee from a cheetah when, A, the gazelle was caught and B, the gazelle escaped. Point C shows the median 1st stott distance. (Data from Caro 1986. Reproduced with permission of Elsevier.)

attacked by two different predators, the sparrowhawk and the peregrine falcon. Sparrowhawks rely heavily upon surprise when they hunt waders, tending to break low from cover and snatch prey from the ground. Peregrines, on the other hand, circle high above potential prey and then stoop towards them at high speed, preferring to seize an aerial target. The redshanks responded to these attack patterns in different ways. Faced with a sparrowhawk they tended to take to the air, but when attacked by a peregrine they would remain still, crouching low against the ground. This shows that the birds were able to rapidly recognize the nature of the threat, assess their risk, and then behave appropriately. This must all happen in an instant and a mistake could be fatal.

Summary

There are a range of primary and secondary strategies that can be adopted by animals to minimize the chances that they will be preyed upon. It should be clear however that no such strategy is likely to be 100% effective and that an element of risk taking may be essential if an animal is to survive. It is unlikely that any one strategy can be 100% effective, and that an animal must constantly reassess the level of danger it faces and react appropriately. Remaining inconspicuous is not always an effective strategy. Often a bold pattern or behavior will deter a predator. There are pros and cons to an individual and group approach to defence, and a trade-off between the costs and benefits of being a group member may result in an optimal group size.

Further reading

In "Interactions between predators and prey" John Endler has written an excellent account of the interrelationships between predators and prey and the coevolution of predatory and antipredatory behavior. This paper can be found in *Behavioural Ecology: An Evolutionary Approach* (ed. by J. Krebs & N. Davies, 1993, Blackwell Science, Oxford), which contains a number of other contributions that should be of interest to any student of animal behavior.

8 Reproductive Behavior: Passing On Your Genes

And God said, be fruitful and multiply, and replenish the Earth.

The book of Genesis chapter 1, verse 28

Contents

With few notable exceptions the means by which the genetic material of the vast majority of multicellular animal species is passed from one generation to the next is sexual reproduction. By this process two animals of different sexes (one male and one female) each contribute a single cell or gamete. These fuse to form a zygote and given the right conditions this develops to become a new and unique genetic individual, but one having genes in

Key points

◆ Male animals produce many, small, cheap gametes whereas females produce few, large, expensive ones. As a consequence of this anisogamy males and females are able to maximize their reproductive effort in different ways.

◆ Anisogamy and imbalances in the operational sex ratio lead to choosiness on the part of the members of the rarer sex and to competition on the part of the more common.

◆ Mate choice may be based on a range of traits that communicate to the chooser something of the reproductive quality of their prospective partner.

◆ A range of possible mating systems exist. The system that is adopted by a given species will depend upon a wide range of evolutionary and environmental factors, but it will often be a compromise between the conflicting priorities of the sexes.

common with both of its parents. The gametes produced by male and female animals, sperm and eggs respectively, are strikingly different. As we will see in this chapter these differences have an important impact upon the reproductive behavior of the animals that produce them.

Males and females are different

An often quoted statistic states that a single human male ejaculate (about 5 ml of fluid containing approximately 350 million sperm) is sufficient to fertilize every woman of reproductive age on the planet today. Of course a statement like this is pretty meaningless in itself but it does serve to highlight an important observation. At a single reproductive event a male produces a very large number of gametes whereas a female produces very few (and often only one). On the face of it this observation may seem to be relatively unimportant, but as we will see this fundamental difference between the sexes has very important implications for the reproductive behavior of animals.

As a consequence of anisogamy a reproductive event may be more "important" to a female animal than it is to a male. For example, when she is in season a lioness will mate with one of the males in her pride every 15 minutes or so. In total she may mate some 3000 times before she finally becomes pregnant. During this time she has contributed a single gamete (the egg) but how many billions of sperm must the male(s) involved have contributed? At a crude level then we might infer that in a pride with just one male (which can happen) the egg has the same relative value as all of the sperm produced during 3000 copulations. So an individual sperm or even a single ejaculate could be considered to have relatively little value to the male in terms of the possibility that it will be the one by which his genetic material is passed on. This situation has important consequences for the social behavior of lions. The small value attached to any particular copulatory event means that in those prides which have a number of males, the males do not compete with one another to monopolize females that are in season. This in turn aids group cohesion and allows the males to devote more of their time and energy to the defence of the pride.

Once fertilization has taken place another important difference between males and females becomes apparent. Assuming that her

> ### Concept
> ### Anisogamy, small cheap males and big costly females
>
> Male animals produce sperm that are small, mobile and cheap to produce. When they are reproductively active they have a very large number of sperm ready for use at any one time. Importantly this reserve can usually be very quickly replenished.
>
> The gametes of female animals (eggs) are, by contrast, relatively large, immobile, expensive to produce, and in finite (short) supply.
>
> This situation is termed anisogamy.

pregnancy continues to the birth of a cub and that the cub survives to independence, the lioness will not be ready to conceive again for more than 2 years (a 3.5-month gestation period followed by a lengthy period during which the cub is dependent upon its mother).

Actual and operational sex ratios differ

Based on his recent performance, however, the male lion will be ready to mate with one of the other females in the pride almost immediately. So in a species where the female provides parental care, females are out of the reproductive "loop" for a considerably longer period than are the noncaring males. If we generalize this observation we find that it holds equally well in those cases where a sex-role reversal takes place such that parental care is in the remit of the male rather than the female. For example female spotted sandpipers (*Actitis macularia*) lay a clutch of four eggs and then in contrast to the majority of bird species abandon their eggs to the care of their male partner. He incubates them and cares for the chicks that hatch from them, and so for that period of time he is not available to mate with another female. However, his "mate" is free to do so and she will often acquire a new male and start another brood of eggs quite quickly. There is actually more to this scenario than meets the eye and we will return to the breeding habits of the spotted sandpiper later in this chapter.

So, depending upon the nature of the reproductive investment that they make there may be an operational imbalance in the ratio of males to females in a population even when there is an actual 1 : 1 sex ratio. As a result the members of the rarer sex will have a heightened value to members of the more common sex and they should compete for access to them.

How do males and females maximize their output?

It is probably safe to assume that individuals of both sexes behave in a way that should maximize their reproductive success in terms of the number and/or quality of young that they are able to produce. Taking into account the consequences of anisogamy that we have thus far discussed, it should be clear that male and female animals could potentially maximize their reproductive output in different ways. Because he can produce a huge supply of cheap gametes a male can most easily increase his output by simply

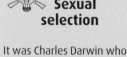

**Concept
Sexual
selection**

It was Charles Darwin who first used the term sexual selection to describe the evolutionary process whereby some individuals gained an advantage over others of their kind in relation to reproduction.

Sexual selection therefore includes both the evolution of competitive advantage in members of the same sex when they compete for the attentions of the other; and the advantage in discriminatory ability exhibited by members of the choosier sex.

fathering as many young as possible, which may mean mating with as many females as possible. Females, on the other hand, cannot usually adopt the same strategy. A female may not be able to increase the number of offspring that she produces beyond a certain level as a result of the size of her gametes, the number of them that are available to her, and the period of time during which she must care for her young. For a female the most effective way by which she can maximize her reproductive output is by maximizing the quality of her offspring. We would therefore expect females to be concerned about the quality of the contribution that the male that she mates with brings to the relationship, either in terms of the quality of his genetic contribution or in terms of the resources that he is able to provide.

It would seem then that if we are to understand reproductive behavior from the male and female position, or at least to explore it effectively, we must consider the behaviors relating to mate choice and those associated with variations in mating systems.

Choosing a mate

It is generally true that females are the "rarer" and therefore choosier sex. We have already considered some of the methods by which males attempt to communicate their undoubted quality to potential mates in Chapter 5 when we examined the evolution of swordtail fish swords and some of the functions of bird song. In the majority of cases it would be true to say that the bigger, louder, showier male is the preferred male and so female guppies *Poecilia reticulata* tend to choose the most brightly colored males available and green tree frogs *Hyla cinerea* select the loudest croaking males. Identifying the trait that is selected does not however tell us what it is that the females are choosing. It is possible that they are simply choosing the males because they have the loudest call or the biggest horns, but it is also possible that their choice is based on some more material benefit that is signaled by the chosen trait. We will consider both of these possibilities in the following sections of this chapter.

Link
Males attempt to communicate their qualities to females.
Chapter 5

Mate choice related to resource provision

We saw in Chapter 5 that female passerine birds use the information transmitted in the courtship song of the male to choose

Plate 8.1 Presumably this Arctic tern male provided a suitable gift! © P. Dunn.

Concept
Extra-pair
copulations

Before the widespread use of DNA analysis to determine paternity it was assumed that the male animal in a socially monogamous pair was the genetic sire of all of the young that he raised.

However analysis of the actual paternity of siblings in a wide range of monogamous species has shown that this is often not the case. Extra-pair copulations or EPCs are relatively common.

We will consider EPCs in more detail later in this chapter.

mates that have good territories. In this way they ensure an adequate supply of food and source of shelter (and perhaps care) for themselves and for the offspring that they will raise. In the case of colonial bird species however the territory might be no more than the few square centimeters of space in which the eggs are laid, and a choosy female might well expect a more tangible demonstration of food availability than the promises contained in a complex song. In the case of the common tern *Sterna hirundo* for example it has been shown that females will mate more often with males that provide them with plenty of courtship gifts in the form of freshly caught fish (Plate 8.1). Jacob González-Solís and his colleagues have shown that food deliveries increase to their maximum rate right around the time that the female is producing her clutch of eggs. Clearly this is when she needs the food the most and so she benefits from having such an attentive mate. However it is likely that the male also benefits from this arrangement, not least one might assume because his chicks will hatch from well-provisioned eggs. But is this a safe assumption to make?

Common terns are socially monogamous (i.e. a pair of birds raise a brood of chicks together), but for long periods of the day the female is alone at the nest while the male is away foraging. In the case of many bird species this situation provides the opportunity for females to take part in extra-pair copulations or EPCs, and this is particularly true in the case of a colonial species such as the common tern where suitable partners are plentiful. It is possible then that a male might unknowingly put effort into provisioning the eggs and chicks of a rival male.

From their observations González-Solís and his team have found that there is a clear correlation between the number of food gifts that a male brings his mate and the number of times they copulate. They also found that the birds copulate far more times than would appear to be necessary for egg fertilization. When we discussed the breeding system of lions earlier in this chapter we saw that frequent copulation with a number of males represented an advantage to the female because this, coupled with her low fertility rate, enabled her to reduce the apparent value of individual copulations to each of her partners. In the case of the terns, however, it is the male that benefits from frequent copulation. Through it he increases the probability that it is his sperm that fertilize her eggs rather than those of his competitors (who we assume had fewer opportunities for copulation). It does seem likely that this strategy is a successful one because relatively few common tern broods contain anything other than full-siblings.

Males of a range of other species provide nuptial gifts as an incentive to females before or during copulation. In the case of the black-tipped hangingfly *Bittacus apicalis* males are under intense pressure to provide females with the right gift. A male presenting an inedible ladybird for example will be spurned from the outset, and even those males who provide a gift that the female is prepared to consider are not always successful. Females of this species decide when to copulate and for how long the joining should last. They will only copulate with a male that provides an edible gift and will only remain with him for as long as it takes to eat it. For successful sperm transfer to take place the flies must copulate for about 5 minutes. If the gift is too small and eaten too quickly the female will leave and the male will have been unsuccessful.

Of course the males of many spider species appear to provide the ultimate in nuptial gifts – themselves. In a species where cannibalism is common mating will always be a risky business, and for this reason the courtship rituals of many spider species are prolonged and complex. But in some species the mating male will literally throw himself into the jaws of his mate during the act of copulation. This would seem to be a maladaptive strategy at first glance. But the fact is that in species such as the Australian red-backed spider (*Latrodectus hasselti*) that perform this behavior males are unlikely to survive the search for a second mate, and females that have recently fed are less likely to encounter a second potential mate than hungry females. So it may be that the only way in which the male can ensure the continuation of his line is through his act of self-sacrifice.

Traits that may communicate a material benefit

A complex song may not be the only trait used by female passerine birds when they choose their mate. For example, when given the choice between two males, a female American goldfinch *Carduelis tristis* will invariably choose the more brightly plumaged of them. Male American goldfinches are strikingly colored. In their breeding plumage they are vivid yellow birds with a striking black cap and an orange beak. The orange and yellow coloration in these birds is the result of the deposition of carotenoid pigments in the feathers and beak as they grow, and so by choosing a brighter male, the females are choosing a mate with high carotenoid levels. But why is that a good thing?

Vertebrates are unable to synthesize carotenoids and so they have to obtain them from their food. The obvious importance of dietary carotenoids has recently emerged as an explanation for a particularly unusual behavior exhibited by the Egyptian vulture *Neophron percnopterus*. These birds are one of only a few species of vertebrate that carry out coprophagy, or to put it another way that eat the feces of other animals, in this case those of ungulates. Ungulate feces are a relatively poor source of food, they are low in both protein (typically less than 5%) and fat (less than 0.5%), and so just why the vultures should expend time and energy on such an unrewarding food has been something of a puzzle. It turns out that they do however contain very high levels of the essential carotenoid lutein. The fact that the birds will go to these lengths to obtain them suggests that carotenoids are a valuable commodity. Interestingly in addition to having unusual dietary habits, Egyptian vultures are also unusual among similar species in that they have a vivid yellow face. This coloration is the result of lutein deposits in the skin and it has been suggested, but not yet confirmed, that this conspicuous patch of color could have a signal function related to mate choice in the same way that the yellow feathers of the American goldfinch do.

So to return to the goldfinches, what information about the male does the female glean from his coloration? Since carotenoids are dietary in origin it could be that the female recognizes the brighter males as better foragers. Given that the males do assist in the rearing of the chicks in this species this could be advantageous, because in a wide range of bird species there is good evidence to support the idea that parents able to provide their young with more food are more successful in terms of the number of their chicks that survive to reproduce themselves. Of course if being a good forager is also a heritable trait the female could be choosing good genes for her offspring. Recently, however, an intriguing new explanation has come to the fore, one that relates carotenoid-based signals to the health status of the individual.

Kevin McGraw and Geoffrey Hill of Auburn University, Alabama have demonstrated that the levels of carotenoids in the beaks and feathers of American goldfinch males that have a heavy infection of endoparasites (specifically coccidians of the genus *Isospora* in their research) were significantly lower than those recorded from unparasitized birds. Similar work involving a variety of species has confirmed the general observation that plumage brightness communicates something about the health status of the male and that females use this information when making their choice of mate. This would be an advantage for a number of reasons. By choosing healthy males females can minimize the risk that they themselves will contract a disease or succumb to a parasitic infection, and of course two healthy parents will be better able to provide for their offspring.

The link between carotenoids and health appears to relate to the role that they play in the functioning of the immune system and in the neutralization of harmful free radicals that, if left unchecked, would damage DNA. Carotenoids stimulate the production of the T and B lymphocytes that are key to the body's antipathogen strategy, and they also play a part in the production of cytokines and interleukins, which in turn play a part in the body's response to injury and inflammation.

It has been suggested by Hamilton and Zuk that elaborate and brightly colored plumage may have evolved in birds as a means by which males might advertise their healthy status to females. Carotenoids, it would appear, could go some way towards strengthening that theory. A sickly male presumably needs to use the carotenoids it obtains from its diet to fight the infections that it succumbs to. A healthy male, by contrast, would not "need" the carotenoids in the same way and could in theory divert them into its plumage as an advertisement of spare capacity and vigor.

Key reference
McGraw, K.J. & Hill, G.E. (2000) Differential effects of endoparasitism on the expression of carotenoid- and melanin-based ornamental coloration. *Proceeding of the Royal Society of London, Series B: Biological Sciences,* **267**, 1525–31.

There has been a tendency for researchers seeking to find evidence that sexual selection can explain the evolution of sexual ornamentation to concentrate upon female choice and male ornaments. Female ornamentation has been largely ignored despite it being far more common than one might at first suspect. By and large studies considering the evolution of female ornaments have tended to concentrate on those species exhibiting a sex-role reversal. So for example it is well known that male pipefish of the genus *Neophis* prefer to mate with females that have conspicuous coloration and elaborate ventral skinfolds. This situation is broadly analogous to female choice in species with conventional sex roles (males being the competitive sex and females the choosier). Recently, however, Trond Amundsen and Elisabet Forsgren have demonstrated that female ornamentation may be influenced by selection acting directly on the females through male choice.

To test their hypothesis that males prefer ornamented females (in this case more colorful) in a non-sex-role-reversed species they made observations of the behavior of male two-spotted gobies *Gobiusculus flavescens* given access to females that varied in the brightness of their belly color. In this species males attract females to their nest sites through an elaborate courtship display that allows them to show off their own iridescent blue fin patterns. The females are not just onlookers in this display, they join in and in their case they arch their bodies to display their round bellies (Plate 8.2). During the breeding season the bellies of these usually dull female fish become yellow or orange because of carotenoids in pigment spots on the skin, and in their eggs that are visible through the skin. As Fig. 8.1 shows males indicated their preference by directing a higher proportion of their courtship displays to more colorful females.

These results strongly suggest that in a species exhibiting conventional sex roles males as well as females may be choosy when it comes to selecting a mate. This may at first seem a little surprising because we have already seen that males are generally expected to be rather indiscriminate in their choice of partner and might be selected to maximize the number of offspring that they produce. But in this particular case this generalization may not apply. There is a limit to the number of eggs a male two-spotted goby can defend in his nest (those belonging to two or three females), and as nests fill up and the breeding season progresses

Plate 8.2 Mutual courtship display of two-spotted gobies. The female (top) displays her colorful belly by bending the body towards the male. © E. Forsgren.

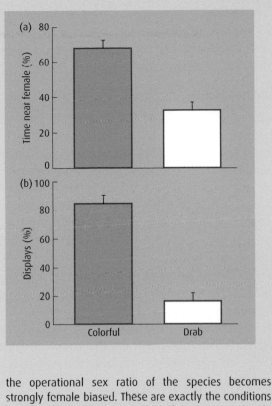

Fig. 8.1 Male two-spotted gobies prefer to associate with colorful females (a) and direct more of their displays towards them (b). (From Amundsen & Forsgren 2001.)

the operational sex ratio of the species becomes strongly female biased. These are exactly the conditions that we might expect to facilitate the evolution of male choosiness.

Amundsen, T. & Forsgren, E. (2001) Male mate choice selects for female coloration in a fish. *Proceedings of the National Academy of Science USA*, **98**, 13155–60.

Choosing traits for their own "attractiveness"

In 1930 R.A. Fisher proposed that females might choose males on the basis of their attractiveness alone, and not because they were good potential parents or gift providers, etc. His theory suggests that initially a particular male trait such as tail or horn length, etc. serves as a means by which females might discriminate between males and identify a preferred type (longer tail for example). The reasons for this choice are not actually important in the case of this theory. There could be some concrete benefit in having a longer tail – it may improve balance or agility and therefore be of assistance in foraging or in predator avoidance. Or it could simply tap into the sensory bias of the females and make the male easier to find or more conspicuous in a crowd. What is important is that

Link
A signal may exploit the sensory biases of its receiver.
Chapter 5

a particular male is chosen and the genes of that particular male and the female that chose him will be passed on to the next generation. That might be the end of the story, were it not for the fact that Fisher imagined a situation in which there was a covariant heritable component to both the trait itself and to the preference exhibited by the female.

So in our example because she chose a long-tailed male a female's sons will inherit their father's tail length and should themselves be more attractive to females (for this reason Fisher's hypothesis is sometimes referred to as the sexy-sons hypothesis). Because of the covariance between the trait and a preference for it, any daughters that are produced will have inherited their mother's preference and so they too will benefit because they are attracted to long-tailed males. So all of the offspring of our original pair possess a reproductive advantage and the future of their genetic lineage should be assured.

This hypothesis is sometimes referred to as runaway selection because the ongoing selective process could result in the evolution of increasingly elaborate traits. Eventually, however, it is likely that a limit will be reached whereby the trait has become so extreme that it has a detrimental effect on the survival of the individual and at that point any further exaggeration of the trait will not be possible. This is the point at which natural selection "applies the brakes."

Being choosy may not always pay off

There are times when it may not pay an individual to be choosy when selecting a mate. Patricia and Allen Moore at the University of Manchester in the UK have found that whilst young female cockroaches will take time over mate selection, an older female mating for the first time will grab the first male that comes along. The researchers found that females that mated first just 6 days after they had attained sexual maturity (at about 100 days old) produced on average 32 young in their first litter and on average 80 young during their 200-day life span. Females that were not allowed to mate until 18 days after they had matured had far lower success, producing just 24 young in their first litter and only 40 during their lifetime. So a delay of just 12 days halved their reproductive output. The behavior of the early and late mating females was also different. Females allowed to mate early were very discriminating about who they mated with, probably basing their choice on the pheromones produced by the male during courtship. But the females that were forced to delay mating for 12 days showed no discrimination and mated very quickly with the first male that they encountered. Presumably falling fertility levels in the aging female result in a heightened motivation to breed (probably through the action of an hormonal cue) that overrides the drive to secure the best available mate.

Case study Horny scarab beetles: an example of an alternative mating strategy

The males of many species are simply unable to compete for access to females on a level playing field. Small male toads, for example, simply cannot croak at the levels needed to attract a female. Instead they adopt a sneaky strategy of hiding close to a larger "croaker" and attempt to steal copulations with females that are attracted to him. Small marine iguanas adopt a more unusual strategy. If they attempt to mount a female a larger more powerful male very quickly displaces them. It takes an iguana approximately 3 minutes to ejaculate and so were it not for a particularly peculiar behavior, small males would never have a chance to inseminate a female. These males circumvent the problem by ejaculating before they attempt to mount the female and storing their sperm in a grove in their penis. In this way even a very short mating might result in successful sperm transfer.

Male scarab beetles *Onthophagus tarsus* exhibit two very distinct morphologies that Armin Moczek and Douglas Emlen suggest might be related to the existence of alternative mating strategies in this species. Larger male scarabs posses a pair of long curved horns that they use as weapons during head-to-head pushing fights. The winners of these fights take possession of breeding tunnels that adult beetles dig directly under the dung piles that will provide food for their developing larvae. Possession of a tunnel is important because tunnels contain females who are ready to breed.

Horned losers quickly move away from the tunnel system of the victor and make no further attempt to gain access to the female within. But not all male scarabs have horns. Nonhorned males are invariably beaten in a contest with a horned male, but instead of fleeing the scene they hang around the tunnel entrance, often hiding beneath the soil. Moczek and Emlen have witnessed these males sneaking into the tunnel system and hiding there when the owner and his mate leave to forage. On her return the intruder male copulated with the female before escaping back up the tunnel system. On other occasions the hornless males sneaked past the horned owner and then ran down the tunnel to mate with the female while the owner, encumbered by his horns, was forced to emerge from the tunnel and then turn around and re-enter in order to give chase. Experiments that compared pairs of animals in an artificial tunnel system have confirmed that a lack of horns is a distinct advantage to a running animal (Fig. 8.2).

So male scarab beetles differ in the behavioral tactics that they use to obtain mating opportunities, they adopt alternative reproductive strategies. The existence of these alternatives has probably favored the evolution of the extreme morphological dimorphism exhibited in this species. Horns are an advantage to bigger males who engage in fighting behavior (Fig. 8.2). But they would not benefit small males who depend upon agility and subterfuge when sneaking in and out of tunnel systems.

Moczek, A.P. & Emlen, D.J. (2000) Male horn dimorphism in the scarab beetle, *Onthophagus tarsus*: do alternative reproductive tactics favor alternative phenotypes? *Animal Behaviour*, 59, 459–66.

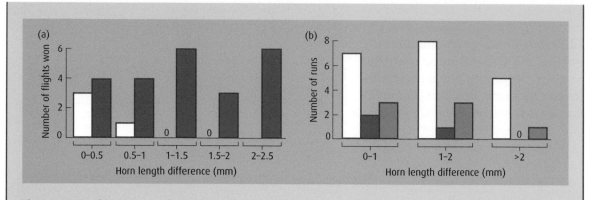

Fig. 8.2 Longer horns are an advantage to a fighting beetle (a), but short-horned males make better runners (b). (a) White bars, shorter-horned winner; blue bars, longer-horned winner. (b) White bars, shorter-horned outperformed longer-horned; purple bars, longer-horned outperformed shorter-horned; lilac bars, no difference in performance between morphs. (From Moczek, A.P. & Emlen, D.J. (2000) Male horn dimorphism in the scarab beetle, *Onthophagus tarsus*: do alternative reproductive tactics favour alternative phenotypes? *Animal Behaviour*, 59, 459–66. Reproduced with permission of Elsevier.)

Mating systems

The number of individuals that they involve generally defines mating systems; i.e. the numbers of mates that an individual has during a mating season. Traditionally such distinctions would have focused upon reproductive social structure (i.e. the numbers of mates involved in the care of young, or the numbers of partners with which an individual was seen to copulate), but recent advances in population genetics have shifted the emphasis so that we now also consider the genetic dimension (the number of individuals contributing genes to the offspring that are raised). For example, we would define social monogamy as a single male and a single female cooperating to raise a litter of offspring. However this would only be an example of genetic monogamy if all of the young in the litter shared genetic material in common with both members of the social pair. In fact as I mentioned above EPCs are common in many socially monogamous species and male animals often unwittingly raise the young of their competitors as their own.

We have already discussed the fact that sexual selection favors the male who mates with a number of females to increase his reproductive output and that at the same time selection favors females that mate with the best males. With this in mind we might predict that polygyny would be the ideal mating system to adopt. After all in a polygynous system males do get to mate with a number of females and each female is probably able to choose to mate with the best male, and if polygyny is ideal why do monogamy and polyandry occur at all? To help us to understand the pros and cons of the various mating systems and to explore the various reasons why the rhetorical question that I have just posed is an unhelpful one I want to briefly consider the mating system of a passerine bird that has been the subject of substantial scientific scrutiny.

The mating system of the dunnock, a sexual conflict

The reproductive behavior of the dunnock *Prunella modularis* has been the subject of a of a body of work by Nick Davies of the University of Cambridge that is quite rightly considered to be a classic of its kind. Dunnocks do not adopt a single mating strategy. Within any one population cases of monogamy (social and genetic), polygyny, polyandry, and polygynandry can all be identified. By investigating the relative outcomes of these different systems in terms of the success enjoyed by the male and female individuals involved, Davies and his team have been able to explain the behavior of the members of the two sexes and to suggest that we should in fact view the

Focus on mating systems

Monogamy

In the case of genetic and social monogamy one male and one female animal are involved in the production of offspring. Genetic monogamy is not a requirement of social monogamy and males are often cuckolded and rear the offspring of their rivals. Monogamy is most common in situations where input from both parents is required to successfully rear the young, or in situations where a male needs to guard his mate from rivals to ensure that he is the genetic father of her offspring.

Resource-based polygyny

When males control access to the resources that are essential to the reproductive success of the females (food, territory, etc.), and males compete with one another to control these resources, polygyny is favored. Under this system a small number of males secure the bulk of the resource and monopolize the majority of the reproductive opportunity. The females choose these high status/quality males and each male will mate with many females.

Female-defence polygyny

This system is similar to resource-based polygyny in that it results in a situation in which a small proportion of the male population establishes reproductive control over the bulk if not all of the female population and high status males secure large numbers of mates. In this case, however, males compete directly with one another to secure and defend groups of females rather than resources.

Lek-based polygyny

Polygyny is also possible when males gather at traditional display sites termed leks. These males do not defend resources (other than a very small area in which they display), nor do they defend females. Females visit the lek site to mate with favored males (a small number of males gain the majority of the matings). The males play no part in the rearing of the young.

Polyandry

Females of some species will take several mates during a breeding season. This system is termed polyandry, and in comparison to the mating systems previously outlined it is comparatively rare.

Polygynandry

As a combination of polygyny and polyandry this mating system involves a number of males and a number of females in a single mating event.

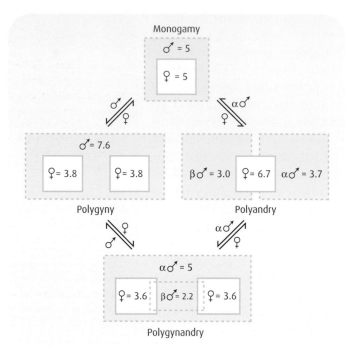

Fig. 8.3 The complicated mating system of the dunnock. See text for a full explanation. The numbers refer to the average reproductive success (number of young raised) per bird in each scenario. The arrows show the directions in which we would expect birds to attempt to shift the system to increase their own output. α birds settle first, β birds join the system later. (From Davies, N.B. (1992) *Dunnock Behaviour and Social Evolution.* Oxford University Press, Oxford. Reprinted by permission of Oxford University Press.)

mating game not as a cooperation between the sexes but as a conflict resulting in a compromise in the interests of one or both of them.

Figure 8.3 shows that in terms of the number of offspring produced, a measure of reproductive success, male birds can do best if they are involved in a polygynous pairing. To achieve polygyny monogamous males will vigorously court additional females. At the same time, however, a monogamous female will attempt to drive away any additional females that her mate does attract. After all if she is forced to share her mate her own reproductive success will be diminished. On the other hand, a monogamous female can benefit from the presence of a second male and so she will solicit copulations from them and attempt to settle into a polyandrous relationship. Clearly this is not to the advantage of the dominant male (her primary mate) and he will attempt to drive away the intruder. Polygynandry benefits both the alpha male of a polyandrous group and the female in a polygynous group and so we would expect both of these individuals to attempt to acquire an additional mate. However, their partners in the original polygynous/polyandrous groups will not benefit from polygynandry and should make attempts to maintain the status quo.

It should be clear then that the actual system adopted by any particular group of dunnocks must represent some form of compromise which must depend just as much on the relative competitive abilities of the individuals concerned and the ratio of males to females in the population as it does upon any motivation to achieve the ideal in sexual selection terms. It will be necessary, therefore, to take into account a very wide range of social, environmental and historical considerations in explaining the evolution and persistence of any one species reproductive behavior.

Some situations do favor monogamy

Although the advent of genetic techniques to determine paternity have significantly altered the notion that socially monogamous animals are truly faithful to one partner, monogamy does exist. It may not be true that swans, for some people the very symbol of fidelity unto death, remain solitary following the demise of their partner, but it is true that for so long as both members of the pair are still alive they do remain faithful to one another.

The environmental conditions faced by a number of species mean that monogamy is the only strategy available to them. For example, because of the extreme Siberian cold that they experience it is unlikely that a female Djungarian hamster *Phodopus campbelii* would be able to keep herself and her litter warm enough to survive. As Fig. 8.4 shows the benefit to the young if their father remains with their mother and they share the

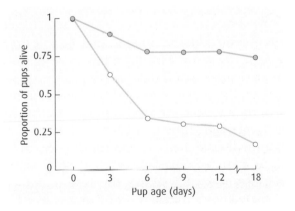

Fig. 8.4 Djungarian hamster litters with both parents in attendance (solid circles) enjoy greater success than those raised by a single mother (open circles). (From Jones, J.S. & Wynne-Edwards, K.E. (2001) Paternal behaviour in biparietal hamsters, *Phodopus campbelli*, does not require contact with the pregnant female. *Animal Behaviour*, **62**, 453–64. Reproduced with permission of Elsevier.)

parental burden is evident, even in the controlled environment of the laboratory. The requirement for shared parental care is probably a very fundamental factor in the stability of the monogamous system in a wide range of species. Interestingly, the male Djungarian hamster takes his role to an unusual extreme, by dragging the emerging young from the birth canal of their mother he acts as her midwife.

Similarly the environment and ecology of some species of deep-sea angler fish dictate that monogamy is essential for their reproductive success. Because they are sit-and-wait predators living in a vast, dark space the chances of males and females meeting are relatively small. Of course the chances of a member of either sex meeting more than one potential mate are even more remote, and so when a pair do meet they stay together – quite literally. When he meets a female, the male angler fish gives up his independent lifestyle and attaches himself as a parasite to his mate taking nutrients directly from her blood.

Mutually beneficial monogamy is, however, a relatively rare scenario and in many cases male animals need to employ physical mechanisms or behavioral strategies to ensure the success of the mating attempts that they are involved in, and the paternity of the young that they raise. One such strategy would be to prevent their mate from taking part in EPCs. For example, earlier in this chapter we saw that through a combination of gift-giving and frequent copulation male common terns can be confident that the chicks that they raise are their own. But if EPCs are beneficial to females we would of course expect them to be just as creative in attempting to stray themselves if it helps them to secure the best possible genetic material for their young.

Mate-guarding

By physically guarding his mate a male may be able to prevent her copulating with his rivals. Monogamous male dunnocks mate-guard for more than 75% of the day when their mate is in season, and during that time they can be quite successful in maintaining the fidelity of their female. As one would predict there is a tendency for the level of guarding to increase to a maximum at about the time that the female ovulates and the chances of fertilization are highest. As a strategy mate-guarding clearly pays off for the dunnock because, as Fig. 8.5 shows, the experimental removal of a territorial male allows intruder males access to his mate, and

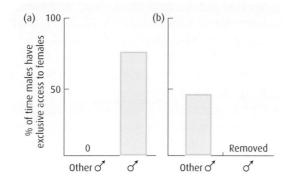

Fig. 8.5 When monogamous male dunnocks guard their females (a) other males can not gain access to them. If the monogamous male is removed they can and do (b). (From Davies, N.B. (1992) *Dunnock Behaviour and Social Evolution.* Oxford University Press, Oxford.)

presumably this provides the female with the opportunity to carry out EPCs.

Some female animals take advantage of male mate-guarding behavior, but do not pay the cost. By which I mean they are not restricted to using the sperm of the guarding male to fertilize their eggs. Female cancrid and portunid crabs can only copulate for a limited period of time following their molt when their shells are still soft. At this time they are particularly vulnerable to predators and to potentially lethal attacks by females of their own species. They therefore benefit from the pre- and post-copulatory guarding with which males provide them. The males, as we have seen, can also expect to benefit from the guarding investment because they will be able to prevent copulations with their rivals. As an added precaution the ejaculate of many crab species solidifies to form a "sperm-plug" within the females' reproductive tract. This plug probably physically prevents fertilization of the eggs by the sperm deposited at subsequent mating events. In the case of *Cancer gracilis* the first male to mate with a female will often desert her after a period of guarding, but before her shell has hardened. Additional males that encounter her in this state will attempt to copulate with her, but they leave almost immediately, providing no further guarding. It is probable that the sperm plug left by her first mate prevents successful copulation with the second. If guarding is valuable to the female we might expect the evolution of a mechanism to prolong it. Such a mechanism may be evident in the case of *Cancer magister*, the Dungeness crab.

The reproductive tract of female Dungeness crabs is unusual in that it includes a **bursa** (a sac) positioned inside the vulva, distal to the spermatheca, and opening into the vagina. Sperm plugs in this species are atypical in that they block only half of the length of the vagina and do not occlude the vulva. In this way the plug does not interfere with intromission and a male crab will presumably have no way of knowing that the female he mates with has already mated and that his effort is wasted. The ejaculate of these second males is diverted into the bursa perhaps fooling the male into "believing" that he has successfully inseminated the female. The outcome for the female is a positive one because these duped males do carry out first-male-type post-copulatory guarding.

Mate interference

Male and female burying beetles of the genus *Nicrophorus* cooperate to bury the corpse of a small bird or rodent on which they rear a brood of larvae. They monitor the food available to their young and if the resources available are insufficient they will commit infanticide. From a human perspective this may seem harsh, but in this way the beetles ensure the survival of at least some of their young. Male burying beetles often attempt to attract a second female to their brood chamber. Obviously if they were successful in doing so they could in theory produce a greater number of offspring. But from the perspective of the resident female the male's success would increase the competition for resources faced by her own young. So when she detects the male sex attractant pheromone released by her mate she attacks him. Research has shown that these attacks dramatically reduce the ability of the male to release his pheromone and so burying beetle monogamy may be forced upon males by females.

However, male burying beetles do sometimes go to greater lengths to find another mate. If they are unable to attract a second female to their crypt, male burying beetles will sometimes desert their mate and first brood in an attempt to find another partner.

Stuart Blackman and his coworkers at the University of Edinburgh have manipulated male desertion behavior in order to further examine the importance of monogamy and male assistance in parental care in these beetles. They allowed pairs of beetles to begin broods and then compared the parental care provided by deserted females rearing the brood alone and those sharing the workload with their still resident mate. Some of the results of these observations are presented in Fig. 8.6, which shows that solitary females invest more time in brood-directed behaviors (feeding by regurgitation and carcass management) than in self-directed behaviors such as resting).

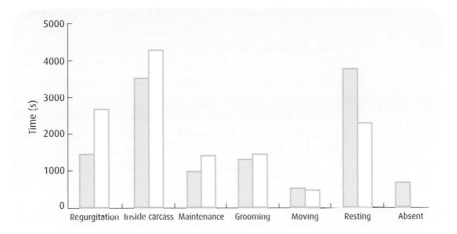

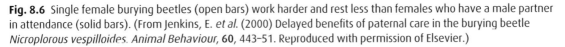

Fig. 8.6 Single female burying beetles (open bars) work harder and rest less than females who have a male partner in attendance (solid bars). (From Jenkins, E. *et al.* (2000) Delayed benefits of paternal care in the burying beetle *Nicroplorous vespilloides. Animal Behaviour*, **60**, 443–51. Reproduced with permission of Elsevier.)

These harder working single mothers may have managed to produce broods of similar quality (number and weight of young), but they paid a real cost in the longer term. When both sets of females were allowed to lay and rear (with no male assistance) a second brood using sperm stored during their copulations with their initial mate the researchers found that previously deserted females did less well. Presumably they had used up important reserves.

This example raises an interesting additional point concerning the evolution of parental care in the burying beetles. Field workers have suggested that in one beetle species close to a third of all the wild broods are produced by a solitary female who will use stored sperm to fertilize her eggs. So even though it may not benefit a male to provide parental care in the shorter term (his efforts seem to do little to enhance the success of the brood he cares for), it may pay in the longer term. This is because it will enhance the success of her next brood and as long as she doesn't mate again these young will be his too.

Sperm precedence and sperm removal

To further increase their paternity assurance male dunnocks engage in frequent copulations and always copulate with their mate if she is known to have spent time with another male, or just

Concept
Sperm competition

Sexual selection, typically thought of as a combination of competition for access to mates by the more common sex, and choosiness on the part of the rarer sex, need not be restricted to the premating period.

At or after copulation males may compete in ways that ensure the dominance of their own sperm over those of other males and sperm themselves may compete with one another for fertilizations. Females may exercise sperm choice by manipulating mating opportunities, by allowing sperm transfer by some partners but not others, by controlling the order of matings, and even through an internal postcopulatory mechanism to select particular sperm to fertilize their eggs.

been out of sight for a period of time. The advantage of this behavior is related to a phenomenon termed last sperm precedence, a consequence of the physical characteristics of the female reproductive tract. Female birds store sperm in tubules positioned at the junction of the vagina and uterus until they are ready to use it to fertilize their eggs. The tubules fill up from the back and so the last sperm to enter are the first sperm to be released. Regular copulation and retaliatory copulation by a male following an EPC probably improve the chance that his sperm will have precedence. Male dunnocks add to this behavior an interesting twist. Prior to copulation the male performs a very unusual display during which he pecks repeatedly at the genital opening or cloaca of the female. This causes the cloaca to become pink and distended and to make regular pumping movements. For a long time the function of this display was unknown although the context was clear and males would not copulate without first going through this ritual. Then after patient observation Nick Davies saw the display end with a sudden dip of the female's abdomen as she discharged a tiny drop of fluid from her cloacal opening. On examination this turned out to be a mass of sperm. Obviously through cloacal pecking the male can thus facilitate the removal of competitor sperm and further improve his own probability of fatherhood.

Sperm removal is a common strategy employed by male invertebrates as a means of securing paternity. For example, the penises of dragonflies and damselflies are covered with bristles and horns that the male uses rather like a scrubbing brush to "clean out" the female's sperm storage vessels before he ejaculates his own sperm into them.

Male bean weevils possess a similarly equipped sex organ, but they use it to a very different effect. Rather than using their spine-covered penis to clean out rival sperm the male bean weevils use it to deter remating by the female. Helen Crudgington and Mike Siva-Jothy of the University of Sheffield in the UK found that the male uses the spines to lacerate the reproductive tract of the female during sex. Their research has shown that whilst a virgin female has a life expectancy of about a month, those females that have had one sexual partner live for only 10 days. With each additional partner that they take their life expectancy shortens and so fidelity is a distinct advantage.

Females should certainly not be thought of as mere bystanders in the reproductive conflict. We have already seen that they make active choices about who to mate with, and female dunnocks do solicit EPCs and can be successful in making a monogamous pairing into a polyandrous relationship. Female bean weevils fight off the amorous advances of unwanted suitors with sharp kicks, thereby limiting the physical damage that they suffer. Through exerting control over the identities and sequences of the partners that they do mate with, females are able to exert control over the identities of the genetic fathers of their young.

Polyandry

The females of some species actively court and mate with a number of males in very rapid succession, but the genetic make-up of the offspring that they produce does not correspond to that predicted by last sperm precedence. By that I mean that the order in which the various males mate with her seems to have no relationship with any one male's likelihood of success. At the same time, however, the genetic make-up of the young is skewed very strongly in favor of one of the males. This suggests that sperm competition is occurring after copulation, i.e. inside the body of the female. One such species is the sand lizard *Lacerta agilis*, the reproductive biology of a Swedish population of which has been the subject of research by Mats Olsson and his colleagues. Olsson found that the more males a female lizard copulated with, the greater her reproductive success (the greater the number of healthy young she produced). Through DNA analysis he was able to demonstrate that not only were the majority of the young sired by just one of the males, but also that the father was the male who's own genetic make-up differed most from that of the female. The implication here is that in some way the female is by some as yet unknown mechanism able to "select" the sperm that she uses from a sample of the wider population in a way that maximizes her success by minimizing the risk of inbreeding.

Whilst making observations of the mating behavior of another species of reptile, the dragon lizard (*Ctenophorus fordi*), Olsson and his coworkers have recorded a male behavior that might go some way towards improving the chances of a male when it comes to sperm competition. They noted that males copulated for up to 60% longer when they had recently seen their partner mate with another male than they did when they had no information about the recent sexual behavior of the female. In this species a longer copulation period results in a larger

Key reference
Olson, M.R., Shine, R., Madsen, T., Gullberg, A. & Tegelstrom, H. (1996) Sperm selection by females. *Nature*, **383**, 585.

ejaculate perhaps increasing the male's chances of fertilizing the female's eggs.

Sex role reversal polyandry

As mentioned earlier in this chapter, female spotted sandpipers, in contrast to the vast majority of bird species, abandon their first clutch of eggs to the care of their mate and seek out a second male with which to lay a second clutch. Effectively they behave in a very "male" way. Female birds arrive on their spring breeding grounds in advance of the males and compete with one another to establish territories. When the males do begin to arrive the females, the "best" of which may attract first one and then a second to be their mate, actively court them.

When they hatch young sandpipers are precocial, by which I mean they are able to look after themselves to some degree. They are mobile, able to thermoregulate, and able to feed themselves on the superabundant mayflies that emerge during the sandpiper's breeding season. This means that a single sandpiper can raise a brood of chicks just as effectively as a pair could. But why is it the female that deserts the male rather than the other way around?

In addition to having precocial young, two further factors probably explain this scenario. Firstly, the sex ratio in this species is slightly male biased. This means that a deserting female will have a good chance of attracting a second mate, but it is likely that a deserting male would not. Secondly, female sandpipers are unable to increase their reproductive output by manipulating the size of the clutch of eggs that they lay in the way that other bird species do. A sandpiper clutch always contains four eggs regardless of the amount of food that is available to the female.

So by abandoning her first mate and laying four more eggs with her second mate, a female spotted sandpiper can double her own reproductive success. But what about the deserted male? He probably has little option but to grin and bear it. After all if he deserts the clutch too the best he could hope for would be to mate with another female, who would lay four eggs and then in all probability she would desert him too. There is some evidence, however, that the female's first partner may in fact enjoy greater success then we expect. DNA analysis has revealed that because the female stores sperm from her initial pairing and may use them to fertilize her second clutch, it may be the second male who cares

Key reference
Oring, L.W. *et al.* (1992) Cockoldry through stored sperm in the sequentially polyandrous spotted sandpiper. *Nature*, **359**, 631–3.

for the chicks of the first. In this way the female makes the best use of the genetic resources available to her.

Polygamy related to defence

The marine iguanas (*Amblyrhynchus cristatus*) of the Galapagos Islands aggregate to form large colonies (often more than 1000 individuals) on the shore adjacent to good feeding grounds. From October to January each year the male iguanas compete to divide the colony into small but fiercely defended territories on which they actively court females. To ensure a high degree of reproductive synchrony copulations in the colony occur over just a few days in January. This reduces the impact of predators upon the emerging young, and reduces the probability that females will accidentally dig up one another's nest sites whilst laying their own eggs.

Through her observations of such a colony Krisztina Trillmich has demonstrated that whilst female iguanas will mate with only one male during a season, males will actively court and attempt to mate with as many females as possible. As Fig. 8.7 shows some males are more successful than others.

Male iguanas do not restrict the movement of females in the colony and each female may make use of the territories of several males. Some they use as resting-territories in which they bask during the day and retreat into crevices at night. Others they use as transit-territories through which they move en route to and from their seaweed feeding grounds. So what makes a male successful?

Before answering this question I want to briefly consider the apparently similar mating system of another coastal breeder. In the case of the elephant seal *Mirounga angustirostris* a male's success is determined by his size and strength. This species, like the

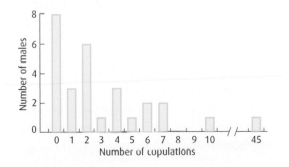

Fig. 8.7 Most male marine iguanas achieve relatively few copulations, but a few males copulate a large number of times. (From Trillmich, K.G.K. (1983) The mating system of the marine iguana (*Amblyrhyncus cristatus*). *Zeitschrift für Tierpsychol*, **63**, 141–72.)

iguana, has a short breeding season during which males and females congregate in huge numbers on suitable areas of the shore. The male animals compete to defend areas of beach in which groups of females first give birth to their young and then very quickly remate. Only the very largest male seals can successfully defend the females on his area of the beach from his competitors. These **beach-masters** as they are known represent a small proportion of the total male seal population, and even then just 4% of them account for more than 85% of the copulations that take place.

So is it size that makes a male iguana successful? Trillmich did find that in general terms female iguanas prefer larger males. But she also found that smaller males on resting-territories copulated with more females than larger males on transit-territories. This suggests that location is important too. The females most often chose to mate with males on higher shore resting-territories. These sites generally consisted of large slabs of flat rock surrounded by jumbles of boulders providing the best basking areas and containing numerous crevices for use as refuges.

In the case of both the seals and the iguanas it is the defensive abilities of the male that enables him to be a polygamist. In the case of the iguana defence of a resource essential to the females (suitable space to bask and hide) enables a male to attract multiple mates. In the case of the seals the male defends the harem of females. For this reason we would describe the former as an example of **resource defence polygyny** and the latter as an example of **female defence polygyny**.

When to be polygamous

Female elephant seals and marine iguanas can afford to be the mates of a polygamous male because in terms of reproductive effort they expect nothing more from the male than good quality sperm with which to fertilize their eggs. But as we have seen there are many cases where a female needs more from her mate. Why then would she apparently choose to share him if doing so means that she will pay a cost as apparently happens in the case of the pied flycatcher (*Ficedula hypoleuca*)?

Typically male pied flycatchers establish a woodland territory to which they attract a female. Then when his primary mate is incubating her eggs he establishes a second territory and attracts a second mate. He abandons this female when her clutch is

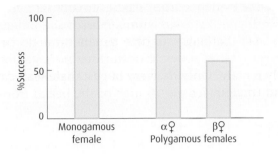

Fig. 8.8 The relative success of monogamous and polygamous pied flycatcher females. (Polygamous female success rates are expressed relative to that of the monogamous female.) (Data from Alatalo, R.V. & Lundberg, A. (1984) Polyterritorial polygyny in the pied flycatcher *Ficedula hypoleuca. Annales Zoologici Fennici*, **21**, 217–28.)

complete and returns to his primary mate to help her to rear her young. The secondary mate receives no assistance when raising her chicks.

It would seem from Fig. 8.8 that whilst the cost to the primary female is probably negligible, the secondary female pays a real cost, raising on average only 60% as many chicks as a monogamous female. From everything that we have discussed so far we should be surprised that the secondary female accepts this situation.

In 1969 Gordon Orians proposed a graphical model to explain the conditions under which this scenario could be explained. His **polygyny threshold model** (Fig. 8.9) suggests that a female should sample the territories of the available mated and unmated males. She should then only choose to be the second mate of an already mated male if by doing so she will achieve greater reproductive success than she would were she to join an unmated male on a poorer territory.

Fig. 8.9 The polygyny threshold model. (a) A female has the choice of settling on territory B (monogamy) or joining the pair on territory A (polygyny). (b) Female reproductive success varies with territory quality. If the difference in quality exceeds PT (the polygyny threshold), a female should choose polygyny, and pay the cost (c) of sharing. (From Krebs, J.R. & Davies, N.B. (1993) *An Introduction to Behavioural Ecology.* Blackwell Science, Oxford, reproduced with permission of Blackwell Publishing Ltd; based on Orions, G.H. (1969) On the evolution of mating systems in birds and mammals. *American Naturalist*, **104**, 589–603. Reproduced with permission of The University of Chicago Press.)

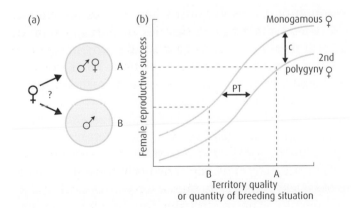

There is some evidence from studies involving populations of great reed warblers *Acrocephalus arundinaceus* and indigo buntings *Passerina cyanea* that suggests that female birds do behave in a way that supports the model. In both cases monogamous females did only marginally better in terms of chicks produced per season than did polygamous females.

Deceitful males

In the case of the pied flycatcher, however, Fig. 8.8 shows that whilst primary polygamous females do almost as well as monogamous females, the secondary females that we would assume are choosing monogamy do considerably less well. Female pied flycatchers it would appear are not therefore behaving in a manner consistent with the polygyny threshold model. In fact it seems highly likely that these females are taken in by deceitful males, and that they are not aware that they are settling down with an already mated individual. Consider the evidence – the male waits until his female is nest-bound before searching for another mate, and he will travel some distance (up to 3.5 kilometers) rather than establish his second territory close to his first. For her part the secondary female seems unable to discriminate between a mated male and an unmated one, even when to do so would clearly benefit her.

Leks

Paul Sherman, writing in the journal *Nature*, has described the lek as nature's version of a singles' bar. I particularly like this colorful analogy because it does describe the scenario very well. A lek is a traditional display area where groups of males aggregate and defend often-tiny territories on which they display. Females visit these sites to select a mate with whom they will copulate. The male plays no further part in the raising of its young and the female does not gain a resource benefit from the male territory.

As a mating system lekking is comparatively rare, but it has evolved in a very wide range of taxa. Although birds and insects account for the majority of known lekking species (about 200 of them), there are mammals, reptiles and fish that lek. A common feature of all leks is that only one or a few males will get to mate with the females. The majority of the displaying animals will have no luck at all. So why do males join leks? Why indeed does lekking happen at all?

Four main theories have been proposed in an attempt to explain leks. The **hot-spot model** suggests that the males congregate at a physical location that is often visited by females, thereby maximizing the chances that they will meet. There is some evidence to support this theory. For example, male roe deer will abandon their forest lek sites if localized environmental disturbance (such as logging) causes females to change their habitual daily routes of travel. These males having abandoned their own display site will move to join the males at a site that still intercepts a female line of travel.

But studies of other species have generated data that do not support the hot-spot model. For example, when Höglund and his team removed the dominant male great snipe from its display site they did not find that the second-best male moved onto the vacant territory as would be expected if it was the site that was important. Instead the lek collapsed, suggesting that it was built around a particularly high quality male, a hot-shot.

So a **hot-shot model** where leks form around an individual could explain the evolution of the lek in some cases. A further prediction that one could make based on the hot-shot model is that males choose to form a lek around a high quality male because doing so will increase their own chances of mating, and it has been shown that in some species males with display territories adjacent to that of the alpha male do have slightly better success than do males on the edge of the lek.

As an alternative explanation the **female-preference model** suggests that the lek evolves in response to the female need to compare a number of potential mates quickly in order to reduce the duration of this stage of the reproductive process or to minimize exposure to danger. After all if females are attracted to displaying males presumably so are their predators. If females do prefer situations that enable them to compare a number of males simultaneously rather than by visiting a number of solitary males in sequence, we would expect a positive correlation between the number of females visiting a lek and lek size. This is exactly what Todd Shelly has recorded in his study of the lekking behavior of a Mediterranean fruit-fly *Ceratitis capitata*.

By establishing equal numbers of artificial leks of six or 36 males in 60 small cages suspended from branches in an area inhabited by the species, Shelly was able to record the visit rates of

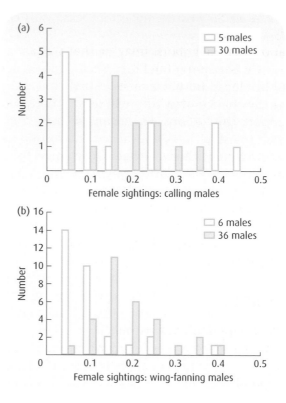

Fig. 8.10 When success is expressed as the ratios of females attending leks to actively courting males (calling (a) or wing-fanning (b)), larger leks do better. (From Shelly, T.E. (2001) Lek size and female visitation in two species of tephritid fruit flies. *Animal Behaviour*, **62**, 33–40. Reproduced with permission of Elsevier.)

female flies to them. As Fig. 8.10 shows larger leks did attract more females than smaller leks. However a similar experiment involving the oriental fruit-fly *Bactrocerus dorsalis* yielded a very different result. In the case of this species lek size did not appear to influence female behavior (Fig. 8.10). This is another indication that a single model is unlikely to explain this intriguing behavior.

The fourth and final hypothesis that I want to consider as a potential explanation of the evolution of leks is the **kin-selection model**. Remember that one of the characteristics of the lek is that few males actually get to mate. We have seen that being close to a hot-shot might increase your chances, but not every male at the site can be close to the hot-shot. The kin-selection hypothesis suggests that even these males may benefit.

Marion Petrie has shown that in captivity male peacocks that have been raised in isolation from one another are more likely to form leks with their genetic relatives (brothers and half brothers) than with more distantly related individuals. Exactly how they recognize one another remains uncertain, but it seems likely that the animals are in some way able to match their own phenotype against that of their neighbors. Imagine that you are a relatively low quality bird. By joining the lek of a high quality relative and adding your display to his you may in fact enhance his success by enabling females to compare him with other males, or by simply making his lek bigger and thereby doing your bit to attract more females.

If the dominant male that you assist in this way is your relative his offspring will be your kin (nephews and nieces), if he is unrelated to you, you will be unrelated to his offspring. So even though a low quality male has little chance of passing on his own genes directly, by lekking with a better brother he might be able to ensure that at least some of his genetic material is transmitted to the next generation.

Key reference
Petrie, M., Krupa, A. & Burke, T. (1999) Peacocks lek with relatives even in the absence of social and environmental cues. *Nature*, **401**, 155–7.

The paradox of the lek

Lek behavior has traditionally been viewed as presenting us with something of a paradox. If all of the females visiting the lek choose to mate with the same male it seems reasonable for us to assume that they all prefer the same male trait. As males without the preferred trait will leave few offspring but males with the trait will produce many "replicas" of themselves, we might also assume that the net result will be a reduction in genetic variation in the preferred trait. In fact the ultimate consequence of this scenario could be a complete loss of variation in the trait in the entire male population in relatively few generations. If this happened we might expect to see a loss of choice in females. If females no longer need to be choosy the lek no longer needs to exist, but of course they do, hence the paradox.

Due to recent advances in the study of genetics a number of plausible explanations for the apparent paradox have presented themselves. A number of researchers working on a range of lekking species have demonstrated that there is in fact a greater degree of genetic variation in sexually selected traits than is the case in nonsexual traits. Perhaps selection for the ongoing development of the trait coincides with selection for increased genetic variation. If the trait is condition-dependent, related to the health of the individual, and condition itself is dependent upon higher levels of genetic variance, then selection for the trait will result in selection for condition and for genetic variance.

As we have seen to be the case in so many of the examples of reproductive behavior that we have considered, it may also be that there is simply more to lekking than we had realized. In a study of the buff-breasted sandpiper *Tryngites subruficollis*, Richard Lanctot and his colleagues have shown that even though relatively few males are responsible for the majority of copulations, the females do mate with more than one male and will use the sperm from more than one partner to fertilize their eggs. In this way the females may maintain genetic variability within the population.

Summary

As a result of differences that exist between them at the gametic level males and females are best able to maximize their reproductive outputs in different ways. Males are driven to aquire as many partners as possible and to compete for access to females. Females, on the other hand, require quality rather than quantity and so should be thought of as the choosier sex. As a result evolution has provided both sexes with an array of strategies to enable them to satisfy their needs. Ultimately, however, the mating system that a species adopts is likely to represent a compromise rather than a victory for either sex.

Questions for discussion

Why should a male spotted sandpiper mate with a female that has already laid one clutch of eggs?

The elaborate tails of male peacocks are assumed to be courtship signals that impress females. Devise a series of experiments that would allow you to establish their effectiveness and to identify what it is about them that females find so attractive.

Further reading

In his book *Dunnock Behavior and Social Evolution* (1992, Oxford University Press, Oxford), Nick Davies explores in amazing detail the complex reproductive system of the dunnock. But for the answers to all of the questions about reproductive behavior that you could ever want to ask read *Promiscuity: An Evolutionary History of Sperm Competition and Sexual Conflict* by Tim Birkhead (2000, Faber & Faber, London).

Index

Page numbers in *italics* refer to figures.

marginal value theorem 135–6
optimal foraging theory 132–7
prey choice 129–37
and risk 140–1
Formica japonica 81
Formica rufa 81
Forsgren, Elisabet 174–5
Fukushi, Tsukasa 80
function of behavior 8

genes 58–66
and migration 90–1
single gene effect 59–61
pleiotropic effect 61–3
polygenic phenotype 63–4
genetic variation 58
genetics 58–9
genotype 58
Gill, E.I. 122
golden lion tamarin 75
González-Solís, Jacob 170
Gould, Carol 150
Gould, James 150
Graham, Paul 81
great tit, camouflage 148
green turtle 77, 87
Greene, Erik 124
Griffen, Andrea 76
group hunting 125–8
guillemot 137, *151*

habituation 68
in *Aplysia californica* 70–4
in crayfish 68–9
in herring gulls 67–8
Hailman, Jack 67
Hanlon, Roger, 146
Hauser, Marc 93
Hawaiian anchovy 126
Helbig, Andreas 90
herring gull *21*
peck reflex 20–1, 67–8
Heteroteuthis sp. 3
Hill, Geoffrey 173
hippocampus 86–7
Hoelzel, Rus 128
homeostasis 43–5
homing behavior 79–82
in pigeons 84–5

honey bee 83
brood cleaning behavior 62, 64–5
horny scarab beetle, mating strategy 177–8
Hughes, R.N. 131, 134
humpback whale 103, 128
hunger 45–6
hyperphagia 45–6

ideal free distribution 138–9, 156
imprinting 17
in cranes 17–18
information transfer hypothesis 125–6
infradian rhythm 50
inking behavior 2–3, 6 8
in blue ringed octopus 8
innate behavior 67–8
involuntary communication 94–5

jack 126, 154
Janik, Vincent 117

killer whale, group hunting 128
kittiwake 12
knee-jerk reflex 21–2
Kruuk, Hans 158

Lanctot, Richard 196
latency 34
lateral giant interneuron (LGI) 38–40
laughing gull 67–8
learning 66–76
associative 69–73
in captivity 75–6
and foraging behavior 121–3
social 74–6
lek 179, 192–6
female preference model 193
hot shot model 193
hot spot model 193
kin selection model 194
leopard seal 154
Leptotyphlops sp. 96
lion, reproductive behavior 167
locus 58
lorikeet 102

Magnhagen, Carin 140
Major, Peter 126
marginal value theorem 137–8

marine iguana
   foraging behavior 129–30
   mating behavior 189–90
marsh tit 87
mate choice 169–76
   and attractiveness 175–6
   costs of 176–7
   female ornamentation 174–5
   and plumage 172–3
   resource provision 169–72
mate-guarding 182–3
mate interference 184–5
mathematical models 133
mating systems 178–96
   dunnock 179–81
   extra-pair copulation 182
   female defence polygyny 179, 190
   lek-based polygyny 179
   monogamy 179, 181–8
   polyandry 179, 187–9
   polygamy 189–92
   polygynandry 179
   resource-based polygyny 179, 190
McGraw, Kevin 178
mechanical communication 108–9
memory
   in *Aplysia californica* 73–4
   in digger wasps 76–8
   landmark 83
   spatial 83–7
menstrual cycle 53
Metcalfe, Neil 45, 153
mimicry 150–1
mobbing 157–9
Moczek, Armin 177
Moore, Allen 176
Moore, Patricia 176
monogamy 179, 181–2, 184
motivation 43–7
mutations 9, 58

nature vs nurture 67
navigation 76–86
   in ants 80–2
   hippocampus 86–7
   in pigeons 84–5
   sun compass 71–2
   in turtles 77, 87, *88*
   using landmarks 82–3

nervous system
   components of 21–6
   control of biological rhythms 53–5
   control of escape behavior 34–40
   control of prey capture 27–34
Neumania papillator 98
neurons 22–3, 34–6
   action potential 24–5
   command neurons 34–5
   giant interneurons 37
   lateral giant interneurons 38–40
   and learning 73–4
   neurotransmitter 22
   resting potential 24
   synapse 22
nuptial gifts 169–72

Octopus cyanea, moving rock strategy 146–7
Olsson, Mats 187
ontogeny 8
optimal diving 137–8
optimal foraging theory 131–7
Orians, Gordon 191
osprey, foraging behavior 124–5
oystercatcher 139

palolo worm 53
Panamanian golden frog, communication 96–7,
   103
Pavlov 70
peacock 194
peregrine falcon 151, 164
pest control 108
Petrie, Marion 194
phenotype 58
pheromones 107
  pest control 108
  trail laying 78–9
phototaxis, turtles 77
pied flycatcher, polygamy 190–1
pigeons 84–5
  homing behavior 71–3
  object location 83
plaice 134
polyandry 179, 180, 187–9
polygamy 189–92
polygyny threshold model 191
prey capture
  nervous control 27–34